MEMBRANE BIOREACTORS

Prepared by the **Membrane Bioreactors** Task Force of the **Water Environment Federation**

Glen T. Daigger, Ph.D., P.E., BCEE, NAE, *Chair*

Adrienne Menniti, Ph.D., *Vice-Chair*

Peter Aerts, Ph.D.
Chibby Alloway
Sara Arabi, Ph.D.
Pierre Bérubé, Ph.D., P.Eng.
Randall S. Booker, Jr., Ph.D., P.E.
Jonathan C. Bundy, P.E.
Ana Calderon
Dave N. Commons
Aileen Condo, P.E.
William J. Conlon, P.E., BCEE, F.ASCE
Nicholas B. Cooper, P.E., BCEE
George V. Crawford
Timur Deniz, Ph.D., P.E.
Jason Diamond, P.Eng.
Ismael Diaz, P.E.
Don Dodson, GA Class 1 WW
Ufuk G. Erdal, Ph.D., P.E.
Val S. Frenkel, Ph.D., P.E., D.WRE
John M. Friel, P.E.
Fred R. Gaines, P.E., BCEE
Andre Gharagozian, P.E.
Anthony D. Greiner, P.E., CCM
Jill M. Hudkins, P.E.
Samuel S. Jeyanayagam, Ph.D., P.E., BCEE
John E. Koch
Terry Krause, P.E., BCEE
Curtis I. Kunihiro, P.E., BCEE
Scott D. Levesque, P.E.
Dennis C. Livingston, P.E.
Jorj Long
Venkatram Mahendraker, Ph.D., P.Eng.
Robert R. McCandless, P.E.
Henryk Melcer
Indra N. Mitra, Ph.D., P.E., MBA, BCEE
JB Neethling, Ph.D., P.E., BCEE
Carsten Owerdieck
Soubhagya Kumar Pattanayak
Jeff Peeters, M.Eng., P.Eng.
Marie-Laure Pellegrin, Ph.D., P.E.
Mark C. Perry, P.E.
Roderick D. Reardon, Jr., P.E.
Gary J. ReVoir II, P.E.
Joel C. Rife, P.E.
Bikram S. Sabherwal, P.E.
Nelson J. Schlater, P.E.
Michael Schmidt, P.E.
Sybil Sharvelle, Ph.D.
H. David Stensel, Ph.D., P.E., BCEE
William Stradling
Vel Subramanian, Ph.D.,P.E., BCEE
Paul M. Sutton
Rudy Tekippe, Ph.D., BCEE, P.E.
Cindy Wallis-Lage, P.E.
Mark Wilf, Ph.D.
Hannah Wilner, P.E.
David S. Wolf, P.E.
Enrique Vadiveloo, P.E.
Don Vandertulip, P.E., BCEE
Seong-Hoon Yoon, Ph.D., P.E.
Thor Young, P.E., BCEE

Under the Direction of the **Municipal Design Subcommittee** of the **Technical Practice Committee**

2011

Water Environment Federation
601 Wythe Street
Alexandria, VA 22314–1994 USA
http://www.wef.org

MEMBRANE BIOREACTORS

WEF Manual of Practice No. 36

Prepared by the Membrane Bioreactors Task Force of the Water Environment Federation

WEF Press

Water Environment Federation Alexandria, Virginia

New York Chicago San Francisco Lisbon London Madrid
Mexico City Milan New Delhi San Juan Seoul
Singapore Sydney Toronto

The McGraw-Hill Companies

Cataloging-in-Publication Data is on file with the Library of Congress.

McGraw-Hill books are available at special quantity discounts to use as premiums and sales promotions, or for use in corporate training programs. To contact a representative please e-mail us at bulksales@mcgraw-hill.com.

Membrane Bioreactors, MOP 36

1 2 3 4 5 6 7 8 9 0 QFR/QFR 1 9 8 7 6 5 4 3 2 1

ISBN 978-0-07-175366-1
MHID 0-07-175366-4

Printed and bound by Quad/Graphics—Fairfield.

This book is printed on acid-free paper.

IMPORTANT NOTICE

The material presented in this publication has been prepared in accordance with generally recognized engineering principles and practices and is for general information only. This information should not be used without first securing competent advice with respect to its suitability for any general or specific application.

The contents of this publication are not intended to be a standard of the Water Environment Federation (WEF) and are not intended for use as a reference in purchase specifications, contracts, regulations, statutes, or any other legal document.

No reference made in this publication to any specific method, product, process, or service constitutes or implies an endorsement, recommendation, or warranty thereof by WEF.

WEF makes no representation or warranty of any kind, whether expressed or implied, concerning the accuracy, product, or process discussed in this publication and assumes no liability.

Anyone using this information assumes all liability arising from such use, including but not limited to infringement of any patent or patents.

About WEF

Formed in 1928, the Water Environment Federation® (WEF®) is a not-for-profit technical and educational organization with members from varied disciplines who work towards WEF's vision to preserve and enhance the global water environment.

For information on membership, publications, and conferences, contact

Water Environment Federation
601 Wythe Street
Alexandria, VA 22314-1994 USA
(703) 684-2400
http://www.wef.org

Manuals of Practice of the Water Environment Federation

The WEF Technical Practice Committee (formerly the Committee on Sewage and Industrial Wastes Practice of the Federation of Sewage and Industrial Wastes Associations) was created by the Federation Board of Control on October 11, 1941. The primary function of the Committee is to originate and produce, through appropriate subcommittees, special publications dealing with technical aspects of the broad interests of the Federation. These publications are intended to provide background information through a review of technical practices and detailed procedures that research and experience have shown to be functional and practical.

Contents

Chapter 1 Introduction

Chapter 2 Membrane Fundamentals

Chapter 3 Biological Process Fundamentals

Chapter 4 Membrane Bioreactor Process Design

Chapter 5 Membrane Bioreactor Facility Design

Chapter 6 Membrane Bioreactor Membrane Equipment Procurement

Chapter 7 Membrane Bioreactor Operation

Appendix A Standard Membrane Bioreactor Computations

Appendix B Glossary

List of Figures

Figure **Page**

List of Tables

Preface

The dramatic increase in the application of membrane bioreactors (MBRs) for wastewater treatment led to significant expansion in knowledge and experience related to their design and operation. Through this expanded experience, practices have evolved which, when routinely applied, lead to successful implementation of MBR technology. The purpose of this Manual of Practice is to summarize and present these practices so that they can be more uniformly used by wastewater professionals. An overview of membrane and biological process fundamentals as they apply to MBRs is first provided and these fundamental principles then support information on the integrated process design of MBR systems. The physical design of features unique to MBRs is addressed and approaches for the procurement of membrane equipment are summarized. Finally, this manual covers operation of properly designed MBR facilities.

This Manual of Practice was produced under the direction of Glen T. Daigger, Ph.D., P.E., BCEE, NAE, *Chair*, and Adrienne Menniti, Ph.D., *Vice-Chair*.

The principal authors of this Manual of Practice are as follows:

Chapter 1	Adrienne Menniti, Ph.D.
	Sara Arabi, Ph.D.
	Glen T. Daigger, Ph.D., P.E., BCEE, NAE
	Fred R. Gaines, P.E., BCEE
Chapter 2	Jill M. Hudkins, P.E.
	Peter Aerts, Ph.D.
	Jonathan C. Bundy, P.E.
	Michael Schmidt, P.E.
	Mark Wilf, Ph.D.
Chapter 3	Ufuk G. Erdal, Ph.D. , P.E.
	Pierre Bérubé, Ph.D., P.Eng.
	Soubhagya Kumar Pattanayak
	Hannah Wilner, P.E.

Chapter 4	Jeff Peeters, M.Eng., P.Eng.
	Ana Calderon, P.Eng.
	Jason Diamond, P.Eng.
	Carsten Owerdieck
	Bikram S. Sabherwal, P.E.
	Cindy Wallis-Lage, P.E.
Chapter 5	Val S. Frenkel, Ph.D., P.E., D.WRE
	Andre Gharagozian, P.E.
	Aileen Kondo, P.E.
	Roderick D. Reardon, Jr., P.E.
	Nelson J. Schlater, P.E.
Chapter 6	Nelson J. Schlater, P.E.
	Val S. Frenkel, Ph.D., P.E., D.WRE
	Jeff Peeters, M.Eng., P.Eng.
Chapter 7	Anthony D. Greiner, P.E., CCM
	Dave N. Commons
	Enrique Vadiveloo, P.E.
	David S. Wolf, P.E.
Appendix A	Adrienne Menniti, Ph.D.
Appendix B	Adrienne Menniti, Ph.D.
	Sara Arabi, Ph.D.

Chapter 2 and Appendix B of this manual were supported by material developed by a task group of the Membranes Community of Practice:

Randall S. Booker, Jr., Ph.D., P.E.
Val S. Frenkel, Ph.D., P.E, D.WRE
Dennis C. Livingston, P.E.
Henryk Melcer
Marie-Laure Pellegrin, Ph.D., P.E.
Paul M. Sutton
Cindy Wallis-Lage, P.E.

Authors' and reviewers' efforts were supported by the following organizations:
AECOM, Atlanta, Georgia, and Orlando, Florida

Black & Veatch Corporation, Alpharetta, Georgia, and Kansas City, Missouri
Brown & Caldwell, Denver, Colorado, and Seattle, Washington
Carollo Engineers, Walnut Creek, California, and Winter Park, Florida
CDM, Albuquerque, New Mexico, and San Antonio, Texas
CH2M Hill, Chicago, Illinois; Columbus, Ohio; Englewood, Colorado; Henderson, Nevada; Portland, Oregon; Santa Ana, California; and Toronto, Ontario, Canada
City of North Las Vegas, Nevada
Colorado State University, Fort Collins, Colorado
Conestoga-Rovers & Associates (CRA), Waterloo, Ontario, Canada
Dewberry, Tulsa, Oklahoma
Dow Water & Process Solutions, Midland, Michigan
GE Water & Process Technologies, Oakville, Ontario, Canada
GHD, Bowie, Maryland
Hazen and Sawyer, P.C., Charlotte, North Carolina, and Hollywood, Florida
HDR Engineering Inc., Folsom, California, and Tampa, Florida
Ion Exchange (India) Ltd.
JJ&G Services, A Jacobs Company, Norcross, Georgia
Jones Edmunds & Associates, Inc., Gainesville, Florida
Kemira Oyj, Helsinki, Finland
Kennedy/Jenks Consultants, San Francisco, California
Koch Membrane Systems, Milwaukee, Wisconsin
Malcolm Pirnie, Atlanta, Georgia
Nalco Company, Naperville, Illinois
Ovivo USA, LLC
PB Americas, Inc., Tampa, Florida
Praxair Inc., Burr Ridge, Illinois
Reiss Engineering, Inc., Orlando, Florida
Short Elliott Hendrickson, Inc., St. Paul, Minnesota
Tetra Tech, Inc., Huntsville, Alabama; Pasadena, California; and Orlando, Florida
University of British Columbia, Vancouver, British Columbia, Canada
University of Washington, Seattle, Washington
Veolia Water, Orinda, California

MEMBRANE BIOREACTORS

Chapter 1

Introduction

1.0 OVERVIEW AND PURPOSE

The use of membrane bioreactors (MBRs) to treat domestic and industrial wastewater has expanded dramatically over the past two decades, leading to significant expansion in knowledge and experience related to their design and operation (Crawford et al., 2000; Judd, 2010). Through this expanded experience, practices have evolved, which, when routinely applied, lead to successful implementation of MBR technology. The purpose of this manual of practice (MOP) is to summarize and present these practices so that they can be more uniformly used by wastewater professionals. Defining current state-of-the-art practices also provides a foundation and basis for further improvements as additional knowledge and experience are accumulated.

Experience also demonstrates that capable wastewater professionals possess much of the knowledge needed to successfully design and operate MBRs, when supplemented by specific knowledge about these systems, especially the membrane component. The principles and procedures routinely used to design and operate the biological component of clarifier-based activated sludge systems are generally applicable to MBRs, as long as the specific conditions and constraints generally incorporated into MBRs are recognized. This basic knowledge must be supplemented by a fundamental understanding of particle removal (i.e., microfiltration and ultrafiltration) membranes to successfully implement MBRs. Thus, the target audience for this MOP is the experienced wastewater professional who is either unfamiliar with the design and operation of MBRs or who needs to enhance his or her knowledge of these systems with current state-of-the-art practices. This MOP should also be helpful to experienced membrane professionals who are interested in expanding their knowledge to include MBRs. Regulators, researchers, and students of wastewater process engineering, facility design, and facility operation will also find this MOP useful.

The principles and practices presented in this MOP are equally applicable to the design and operation of MBRs to treat domestic and industrial wastewater. The characteristics of industrial wastewater can vary widely, especially compared to domestic wastewater, and no attempt is made in this MOP to cover the wide range of industry-specific applications. Membrane bioreactors have been applied to a wide range of applications, ranging from "package" facilities treating the wastewater from individual buildings up to major centralized wastewater treatment plants (WWTPs). The principles presented in this MOP are relevant to all applications, but the specific practices are more focused on engineered systems rather than complete-package systems as might be purchased from a membrane system vendor for the smaller, building-scale applications.

2.0 ORGANIZATION AND STRUCTURE OF THE MANUAL

The MOP consists of seven chapters. This chapter introduces the purpose and scope of the MOP, provides an overview of typical components of an MBR, and introduces standard terminology used throughout the MOP. A brief history of the development of MBR technology and an overview of MBR capabilities and typical drivers for their use are also presented.

Chapter 2 provides an overview of key membrane process fundamentals, including a summary of ancillary systems such as those needed for basic operation of the membrane system and membrane cleaning. This section should be useful for wastewater professionals who are not familiar with the components of the particle separation membranes used in MBRs.

Chapter 3 provides an overview of biological process fundamentals, especially those that apply specifically to MBRs. This chapter will be useful to wastewater treatment professionals seeking to understand how their existing knowledge of biological treatment systems applies to MBRs. It will also help introduce membrane treatment professionals to some of the key concepts needed to design the biological component of an MBR.

Building on the fundamental principles presented in the preceding three chapters, Chapter 4 presents an integrated methodology to develop a complete, integrated process design for an MBR. Chapter 5 addresses the design of the physical facilities needed to implement the MBR process, with emphasis on features unique to MBRs.

Membrane bioreactors present unique challenges to the designer as the membrane system component is critical to system design but each vendor offers systems with unique features. System design must recognize and accommodate these unique features, which results in challenges in integrating membrane equipment procurement into the MBR design procedure. Chapter 6 presents approaches that have been used successfully to address this issue, resulting in successful procurement of necessary membrane equipment and systems.

Finally, building on fundamental and practical information presented in the previous chapters, Chapter 7 outlines procedures and practices leading to the successful operation of properly designed MBR facilities. Topics such as biological and membrane process operation procedures, routine monitoring, facility maintenance, and troubleshooting are addressed.

3.0 MEMBRANE BIOREACTOR PROCESS BACKGROUND

3.1 Membrane Bioreactor Process Introduction

An *MBR* is a combination of suspended-growth activated sludge biological treatment and membrane filtration equipment performing the critical solids–liquid separation function that is typically accomplished using secondary clarifiers. Any biological treatment process relies on microorganisms to metabolize nutrients and organic matter, removing them from the wastewater and converting them to new biomass. In an activated sludge process, the microorganisms aggregate into flocs and these flocs are suspended in the wastewater to facilitate treatment. Once the wastewater is treated, the flocculated microorganisms must be separated from the clean water. Conventionally, a clarifier is used for solid–liquid separation, as shown in Figure 1.1(a). Therefore, successful treatment in a conventional activated sludge process relies on the development of flocs that settle well. A membrane bioreactor simply replaces the clarifier with a membrane for solid–liquid separation, as illustrated in

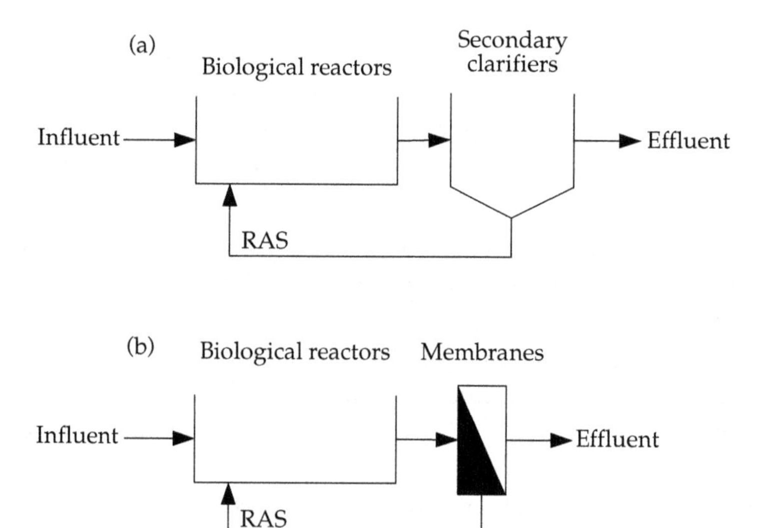

FIGURE 1.1 (a) Conventional activated sludge process with clarifiers for solid–liquid separation; (b) membrane bioreactor process with membrane filtration for solid–liquid separation.

Figure 1.1(b). The concentrated biologically active solids that are separated from the treated wastewater are called *return activated sludge* (RAS). Return activated sludge is returned to the bioreactor and mixed with wastewater being treated. The mixture of the biologically active solids and the wastewater being treated is referred to as *mixed liquor* or *mixed liquor suspended solids* (MLSS).

Conventionally, filtration in water treatment is defined as the separation and removal of solid particles from a liquid (Metcalf and Eddy, 2003). Membrane filtration extends the compounds removed to include dissolved substances (Cheryan, 1998). Membrane filtration processes are often classified by the size of compounds removed from solution. The microfiltration and ultrafiltration membranes used in MBRs remove particles, colloids, and macromolecules based on size exclusion. Microfiltration membranes typically applied in MBR systems have pore sizes of approximately 0.1 to 0.4 μm, whereas ultrafiltration MBR membranes have smaller pore sizes in the range of 0.01 to 0.05 μm. It should be noted that the distinction between these two types of membranes is based largely on the evolution of microfiltration and ultrafiltration membranes specifically applied to MBRs. Broader discussions of this topic can be found in literature by Metcalf and Eddy (2003) and Cheryan (1998).

Pressure is the driving force generating the flow of water through microfiltration and ultrafiltration membranes. The flowrate of water passing through the membrane is often normalized by the membrane area; this membrane performance parameter is called the *membrane flux*. The achievable flux through the membrane during typical operation is a driving factor determining the required membrane area for an MBR and, therefore, is a critical parameter determining the capital cost of the system. Over time during membrane operation, material from the water being treated gradually accumulates on the membrane surface and in the membrane pores, inhibiting the passage of water through the membrane. This accumulation of material, called *membrane fouling*, limits the productivity of the membrane and is detrimental to process performance and membrane life. A significant fraction of the operational cost of MBR systems goes toward controlling or removing membrane fouling. Although MBRs eliminate the need for well settling sludge, it is important to remember that the performance of the membrane system is still directly related to the properties of the MBR biomass. The constituents contained in the mixed liquor (activated sludge flocs, fibers, grease, colloidal and soluble materials, etc.) are responsible for membrane fouling in MBR systems.

3.2 Significant Components of a Membrane Bioreactor System

Figure 1.2 shows typical significant process components of an MBR WWTP. A general description of each major unit process is also provided in this section. Further details on MBR-specific implications and process considerations of each major unit process can be found in subsequent chapters of this MOP. Chapter 4, in particular, provides greater detail on MBR-specific implications and design considerations of preliminary treatment, primary treatment, and fine screening. For details on the general design of the major unit processes that are not specific to MBR systems, refer to *Design of Municipal Wastewater Treatment Plants* (WEF et al., 2009).

3.2.1 *Preliminary Treatment*

Preliminary treatment includes coarse and fine screening; grit removal; and fats, oils, and grease (FOG) removal. Coarse screens are applied to remove large objects and debris that could damage downstream equipment or block flow in pipes and channels, whereas fine screens are applied to remove hair and other biologically inert fibrous material. Without fine screens, hair and other stringy materials build up in membrane modules over time, accelerating membrane fouling and possibly damaging the membranes. Therefore, fine screens extend membrane life and are typically required by MBR vendors. The location of the screen depends on the specific MBR application under design; Chapter 4 provides further discussion on this topic.

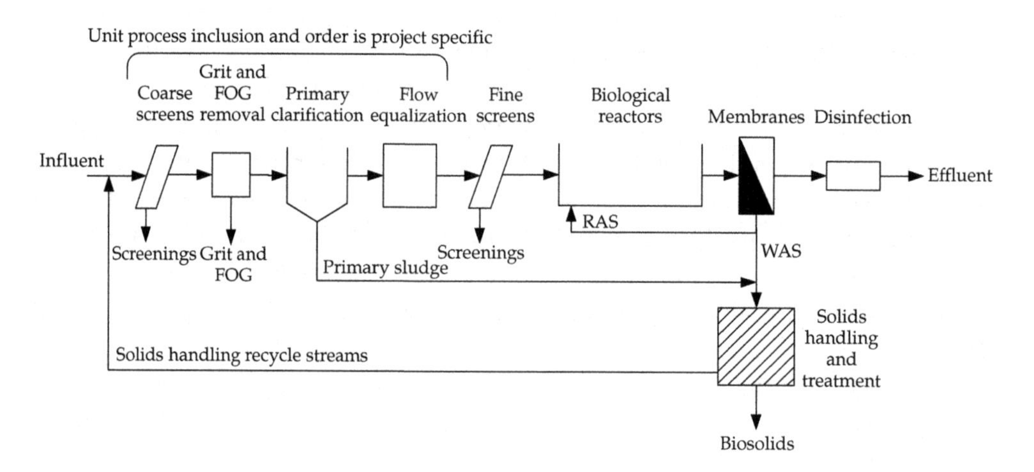

FIGURE 1.2 Overall treatment process flow diagram for an MBR system.

Without specific grit removal designed into the treatment train, grit can accumulate in pipelines and channels, sludge digesters, and bioreactors. Grit is also abrasive and can damage MBR membranes if excessive amounts of grit accumulate in the membrane tank. Therefore, MBR suppliers may require grit removal as part of the membrane system warranty. The concentration of influent FOG is also an important consideration for MBR design. If a treatment plant has excessive amounts of influent FOG, this FOG can accumulate on the membrane surface and accelerate membrane fouling. Furthermore, the presence of FOG in the bioreactor can promote the growth of microorganisms that cause foaming, making process operation difficult. Chapter 4 provides further information on the impacts and design considerations of grit and FOG in MBRs.

3.2.2 *Primary Treatment*

Primary treatment, if incorporated to an MBR design, involves the separation and removal of floating material including FOG and settleable solids using sedimentation. Floating materials are skimmed off the top of the primary clarifier and settled solids are withdrawn from the bottom. Primary treatment is not a requirement for MBR systems and the decision to include the unit process or not is similar to that for conventional systems. Elimination of primary treatment reduces the solids handling requirements and land area associated with the WWTP, whereas inclusion of primary treatment minimizes the oxygen requirements in the bioreactor. Chapter 4 provides further discussion of the impacts and design considerations of primary treatment in MBRs.

3.2.3 *Flow Equalization*

Peak flow treatment is a critical consideration for MBR systems. In conventional activated sludge systems with clarifiers for solid–liquid separation, flowrates in excess of the design capacity can typically be passed through the clarifier at the expense of reduced treatment efficiency and possible solids losses into the effluent. In MBRs, however, the membranes are a hydraulic bottleneck and, for a given membrane area, there is a fixed amount of flow that can be passed through the system. The peak flowrate in MBR systems often dictates the membrane area, particularly in MBR systems subject to significant variation in influent flowrate. Because of the high capital costs of membranes, designing an MBR system with a membrane area sufficient to treat all possible flowrates may not be economically favorable if the peaking factors at the treatment plant are high. Flows in excess of the peak design flow can be equalized externally in an upstream tank or internally within the freeboard volume of the bioreactor. Chapters 4 and 5 provide further discussion on the impacts and design considerations of flow equalization in MBRs.

3.2.4 *Biological Reactors*

Bioreactor design for an MBR system is similar to that of a conventional activated sludge process. However, special care must be taken designing bioreactors for MBR systems with regard to oxygen transfer efficiency. Because of a higher MLSS concentration, the oxygen uptake rate in MBRs is higher than in conventional systems. Typically, high MLSS concentrations decrease the transfer efficiency of aeration systems. Thus, the ability to transfer sufficient oxygen is at times the limiting factor dictating bioreactor size. Special attention should be given to the type of the aeration system and the depth of the bioreactor when designing a bioreactor for an MBR system to ensure that sufficient oxygen can be provided for the biological process. Chapters 3 and 4 provide further information on important design considerations for bioreactor design in MBRs.

3.2.5 *Membrane Systems*

The membrane separation process also includes ancillary systems required to maintain flux through the membranes. These ancillary systems include air scour along the membrane surface to control the deposition of foulants on the membrane. Air scour is a significant source of energy consumption for MBR systems and is a significant factor contributing to the higher energy use in MBRs compared to conventional systems. Membrane equipment systems for MBRs include chemical systems for membrane cleaning; however, chemical cleaning also increases the operating costs in the MBRs. Chapters 2, 4, and 5 provide further information on important design considerations for membrane system design in MBRs.

3.2.6 *Disinfection*

By virtue of size exclusion, membranes remove a significant portion of effluent pathogens compared to conventional activated sludge systems. This action provides a physical mechanism for disinfection. Furthermore, low turbidity and low organic matter in the MBR effluent reduce downstream disinfection requirements. The low turbidity improves the efficiency of UV disinfection, reducing power requirements, while less concentrations of organic matter reduce chlorine demand so less chlorine is required to achieve target residual concentrations.

3.2.7 *Solids Handling and Disposal*

Waste activated sludge handling and disposal for MBRs are not significantly different from conventional wastewater treatment systems. For decentralized or satellite systems, solids are sometimes wasted back to the collection system and conveyed to a larger, centralized treatment facility for regional solids handling and treatment. One unique solids handling option developed through the application of MBR technology is the use of membrane thickening to thicken waste activated sludge before digestion.

3.3 Scales of Membrane Bioreactor Application

Membrane systems have modular designs allowing MBR technology to be successfully applied on a wide range of scales. The smallest MBRs are designed to treat flows less than 378 m^3/d (100 000 gpd) and can be found in places such as office buildings, hotels, or subdivisions, where water can be reused near the source, or in state and national parks where there is no access to a municipal sewer connection. Membrane bioreactor technology is particularly well suited to small treatment systems because MBRs are highly automated and provide reliable treatment. Moreover, MBRs can easily be operated remotely and require minimal daily operator intervention.

In a survey of worldwide MBR installations, Oppenheimer et al. (in development) estimated that the global number of MBRs probably exceeds 5000 installations. The authors also determined that only 166 of these MBR installations have a design capacity greater than 3.8 ML/d (1 mgd). Thus, the most common MBR installations have capacities less than 3.8 ML/d (1 mgd). For installations larger than 3.8 ML/d (1 mgd), Figure 1.3 (taken from Oppenheimer et al. [in development]) demonstrates that the most common treatment size is 3.8 to 19 ML/d (1 to 5 mgd). However, the number of MBR installations in all size ranges continues to grow and, as membrane costs decrease and confidence in the technology builds, MBRs with capacities greater than 38 ML/d (10 mgd) will become more common.

Although there are currently only a handful of MBR facilities with design capacities greater than 19 ML/d (5 mgd), there are drivers behind the use of MBRs on a larger scale. Membrane bioreactor technology may offer a cost-competitive treatment solution for site-constrained facilities that must expand or upgrade using small-footprint treatment processes that can provide a high-level nutrient reduction

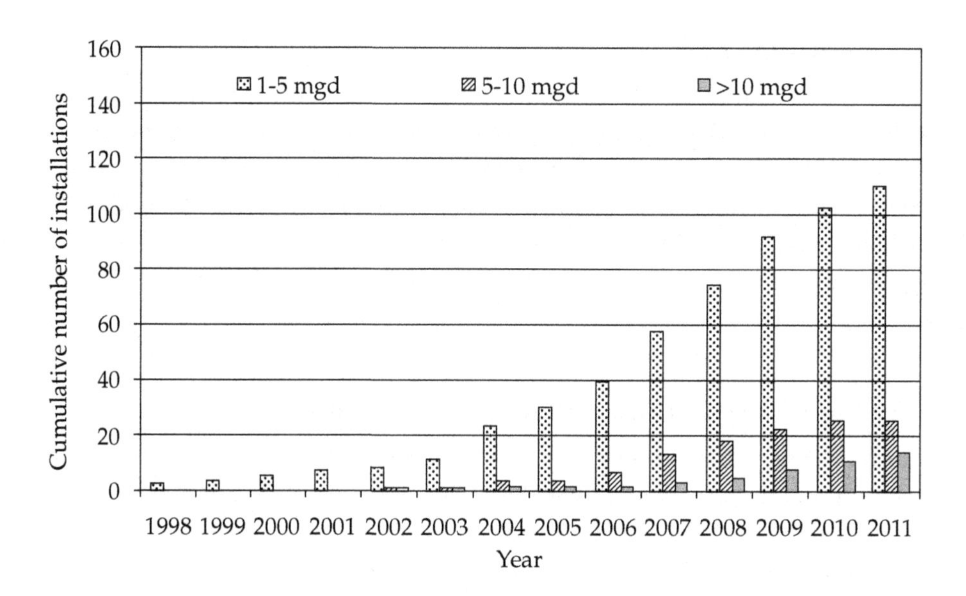

FIGURE 1.3 Cumulative number of MBR installations by capacity (mgd × 3.785 = ML/d) (reprinted, by permission, from Oppenheimer et al., *Investigation of Membrane Bioreactor Effluent Water Quality and Technology.* © 2011 by the WateReuse Research Foundation).

and/or high-quality effluent for water reuse. As of February 2009, Oppenheimer et al. (in development) found that the largest operating North American MBR, located in Tempe, Arizona, has a design capacity of 34 ML/d (9 mgd). In addition, the largest North American MBR under construction is the Brightwater treatment plant in King County, Washington, which has a design capacity of 148 ML/d (39 mgd). Worldwide, even larger MBR installations are currently in operation, under construction, or being designed.

4.0 HISTORY OF MEMBRANE BIOREACTOR DEVELOPMENT

As reported by Singh (1989), in a review of membrane research and development, the work of Sourington and Loeb at the University of California, Los Angeles, on asymmetric, thin-skin, and cellulose acetate membranes is credited with introducing the modern era of membrane use. The application of membrane technology in the water pollution control field included several projects implemented by Dorr-Oliver, Inc. (Milford, Connecticut) in the early 1960s for the separation of fine particulate from waste. In 1969, U.S. Patent 3,472,765 was issued to William E. Budd and Robert W. Okey of Dorr-Oliver for a process that integrated complete-mix activated sludge process with membrane technology.

In 1974, Dorr-Oliver engineers entered into a licensing agreement with Sanki Engineering Company (Tokyo, Japan) (Mishra et al., 1994). The Japanese government had mandated that wastewater is reclaimed and reused to reduce the need for potable water as well as the amount of wastewater being discharged to wastewater treatment works. Sanki installed approximately 20 membrane wastewater reclamation systems similar to the one used in the Pikes Peak visitors' center. These facilities incorporated modules that held cartridges of flat-plate membranes installed in a horizontal plane.

The process, referred to as the *membrane wastewater treatment system*, consisted of raw wastewater being passed through a coarse screen and discharged to an aerated conditioning tank; the flow was then pumped to an aeration tank where it was mixed with RAS. This mixture was then pumped to the membrane water recovery system where the reclaimed wastewater was drawn off and the RAS returned to the aeration tank. According to Dorr-Oliver, operating costs for the systems, including power, 3-year membrane replacement, labor, and sludge wasting and chemicals, was forecast at \$2.61/3.8 m^3 (1000 gal) in 1994 U.S. dollars (Mishra et al., 1994).

The Dorr-Oliver process patent expired in the 1980s. After the patent expired, the company began concentrating on other applications of biological treatment using fluid bed reactors.

At the same time, Thetford Systems (Ann Arbor, Michigan), which later became part of Zenon Environmental (now GE Water & Process Technologies, Trevose, Pennsylvania), manufactured and marketed an on-site treatment and recycling system called Cycle-Let® to meet the need for treating waste from small commercial facilities that required a high-quality effluent. It incorporated an aerobic-anoxic activated sludge process with tubular ultrafiltration in a two-pump feed and bleed loop. Thetford Systems installed 27 Cycle-Let® systems between 1974 and 1982 (Benedek and Cote, 2003); many of these systems are still operating today, including one at a manufacturing facility in New Jersey where the wastewater is reclaimed and used for flushing toilets and watering lawns (Gaines et al., 2005).

Following the first introduction of early MBRs, several factors have been introduced that have led to their increased application. These include

- The development of better materials of construction that increased membrane life;
- Requirements for certification and better training of WWTP operators;
- Application of sophisticated instrumentation and control systems to increase efficiency; and
- More stringent wastewater discharge criteria requiring more efficient process technology.

A combination of work done during the 1970s and 1980s, development of less expensive and more reliable membrane systems, the need for dischargers to comply with more stringent standards, and increasing emphasis on the beneficial reuse of treated wastewater have had a significant effect on the application of membrane technology. Systems that were once much more expensive than conventional technology being applied were now competitive from a cost standpoint. One of the results was that the early market leaders such as Zenon Environmental and Kubota Corporation, Tokyo, Japan (http://www.kubota-mbr.com/), saw competition from many other firms.

Some of these manufacturers offered membranes for incorporation in systems designed by other vendors or independent design firms while some offered complete systems. As the market has expanded, the proprietary nature of the technology has become less of a mystery and experienced designers are now able to design their own systems. Membrane types vary, including various configurations of hollow-fiber and flat-plate systems. As manufacturers, engineers, and plant operations personnel gain experience in the design, construction, and operation of MBR systems and MBR total costs are reduced, their acceptance and use is expected to continue to grow.

5.0 OVERVIEW OF MEMBRANE BIOREACTOR CAPABILITIES

5.1 Benefits of Membrane Bioreactor Technology

5.1.1 Footprint

In terms of carbon and nutrient removal, MBRs can accomplish the same level of biological treatment based on the same fundamental biological principles as conventional activated sludge systems. However, MBRs accomplish this in a smaller footprint because a higher mixed liquor concentration can be maintained in the aeration basins and because membranes require less space than clarifiers. For a constrained treatment plant site, this allows capacity expansion in the same treatment plant footprint by retrofitting a conventional activated sludge system to an MBR. For a green field site, MBRs can accomplish the same capacity of treatment in a smaller footprint compared to conventional activated sludge systems.

5.1.2 *Absolute Biomass Retention and Ease of Operation*

The membrane in an MBR provides complete biomass retention; therefore, biomass washout due to poor settling sludge is not a concern. Because the activated sludge properties are decoupled from biomass retention and level of treatment, MBRs can be highly automated. This high level of automation allows complete control of the solids retention time (SRT) and allows MBRs to be applied in situations where operator attention needs to be minimized, such as with decentralized or remote treatment systems.

5.1.3 *Effluent Quality*

The membrane also produces a higher level of treatment than clarified secondary effluent coupled with media filtration. The small pore size of the membrane retains all suspended solids along with some colloidal and macromolecular compounds. This high degree of filtration removes not only the carbon, but also the organic nutrients associated with it. Therefore, MBRs (and membranes, in general) provide benefits for treatment plants producing reuse water or with stringent effluent discharge limits for suspended solids, organic matter, total phosphorus, or total nitrogen. Biological nutrient removal can be accomplished in the activated sludge process and effluent polishing is provided by the membrane. The higher effluent quality also reduces disinfection requirements.

5.2 Challenges of Membrane Bioreactor Technology

5.2.1 *Cost*

The comparative costs between MBRs and other treatment alternatives depend largely on the goals of the project being evaluated. All the unit processes required to accomplish a given treatment objective should be included when calculating capital and operating costs and when making process decisions. The following analysis of MBR costs is based on a comparison to conventional activated sludge systems.

Compared to conventional activated sludge systems, membrane bioreactor systems typically require a higher level of pretreatment to protect membranes and avoid excessive fouling, regardless of whether primary clarification is eliminated or not. There are added capital and operational costs associated with these additional unit processes. Furthermore, membranes are expensive and create a high initial capital investment; there is also a long-term cost for membrane replacement. Using air scour to control membrane fouling is energy-intensive and chemicals are also required for membrane cleaning. Therefore, the operational costs of MBRs may be higher than conventional activated sludge systems.

5.2.2 *Capacity*

As discussed previously, MBRs have a limited peak flow capacity, which may require upstream flow equalization or additional freeboard within the aeration basin to generate a cost-effective MBR design. Providing multiple process tanks and extra membrane trains can provide flexibility to meet all capacity requirements, but may result in higher capital costs. Furthermore, treatment plants dealing with high wet weather flows may need to develop unique solutions to accomplish cost-effective treatment of all plant flows.

5.2.3 *New Technology*

Membrane equipment systems are unique to each vendor; therefore, preselection or prepurchase is required to ensure that membrane systems are designed to adequately address the characteristics of the membrane system chosen. Furthermore, MBR technology is still relatively new and, thus, the life expectancy of membranes is still somewhat unknown, even for the largest, most established vendors.

5.3 Typical Drivers for Membrane Bioreactor Installation

Given the set of benefits and challenges for MBR technology, there are a number of drivers for MBR installation. Typical MBR drivers are as follows:

- Small treatment plant footprint required to achieve a given capacity;
- Capacity expansion must occur without significant added reactor volume or overall treatment plant footprint;
- High-quality effluent required for water reuse applications;
- Treatment plant has stringent discharge limits for suspended solids, organic matter, total nitrogen, or total phosphorus;

- Decentralized wastewater treatment or water reuse strategies create the need for unstaffed operation;
- The application calls for a highly automated system; and
- Elimination of biomass washout and/or absolute SRT control is desired.

6.0 REFERENCES

Benedek, A.; Cote, P. (2003) *Long Term Experience with Hollow Fibre Membrane Bioreactors;* BAH03-180; International Desalination Association: Topsfield, Massachusetts.

Cheryan, M. (1998) *The Microfiltration and Ultrafiltration Handbook*; Technomic Publishing: Lancaster, Pennsylvania.

Crawford, G.; Thompson, D.; Lozier, J.; Daigger, G.; Fleischer, E. (2000) Membrane Bioreactors—A Designer's Perspective. *Proceedings of the 73rd Annual Water Environment Federation Technical Exposition and Conference* [CD-ROM]; Anaheim, California, Oct 14–18; Water Environment Federation: Alexandria, Virginia.

Gaines, F. R.; Marshall, J.; Halloran, D. (2005) Wastewater Reuse from Onsite Systems. *Proceeding of the 78th Annual Water Environment Federation Technical Exhibition and Conference* [CD-ROM]; Washington, D.C., Oct 29–Nov 2; Water Environment Federation: Alexandria, Virginia.

Judd, S. (2010) *The MBR Book, Second Edition; Principles and Applications of Membrane Bioreactors for Water and Wastewater Treatment;* Butterworth-Heineman: London, U.K.

Metcalf and Eddy, Inc. (2003) *Wastewater Engineering: Treatment and Reuse,* 4th ed.; McGraw-Hill: New York.

Mishra, P. N.; Sutton, P. M.; Mourato, D. (1994) Innovative Liquid Phase Treatment Systems: Combining Membrane Technology and Biological Processes. *Proceeding of the 67th Annual Water Environment Federation Technical Exposition and Conference;* Chicago, Illinois, Oct 15–19; Water Environment Federation: Alexandria, Virginia.

Singh, R. (1989) Surface Properties in Membrane Filtration. *Chem. Eng. Prog.*, **85** (5/6), 59–64.

Oppenheimer, J.; Rittmann, B.; DeCarolis, J.; Hirani, Z.; Kiser, A. (in development) *Investigation of Membrane Bioreactor Effluent Water Quality and Technology*; Report Number WRF-06-007; WateReuse Research Foundation: Alexandria, Virginia.

Water Environment Federation; American Society of Civil Engineers; Environmental and Water Resources Institute (2009) *Design of Municipal Wastewater Treatment Plants*, 5th ed.; WEF Manual of Practice No. 8; ASCE Manual and Report on Engineering Practice No. 76; McGraw-Hill: New York.

Chapter 2

Membrane Fundamentals

(continued)

1.0 INTRODUCTION

This chapter provides an overview of key process membrane fundamentals, including a review of membrane processes and membrane bioreactor (MBR) systems. The first section reviews membrane process fundamentals to provide an understanding of commonly used terms to define membrane operations. The remaining sections include a review of membrane processes incorporated to MBRs, membrane process configurations, system components for MBRs, and membrane operation and maintenance (O&M) fundamentals. These concepts are introduced in this chapter; more detailed information is provided in subsequent chapters.

Membranes are used to remove a variety of constituents from water and wastewater including suspended solids, colloids, and dissolved particles. The type of membrane used is dependent on the constituent(s) to be removed. Membranes are classified based on pore size and, therefore, the size of the particle or solute to be removed. Membrane classifications and characteristics are discussed in detail in Section 2.0 of this chapter.

1.1 Membrane Bioreactor Process Overview

An overview of membrane bioreactors was provided in Chapter 1, including an overall treatment process flow diagram in Figure 1.1. A variety of ancillary systems, some of which are shown in Figure 2.1, are associated with the membrane system. These may include the following: biological process blowers; membrane scour blowers; recirculation pumps; permeate pumps; backpulse and clean-in-place (CIP) tanks; backpulse and CIP pumps; air compressors; and dip tanks.

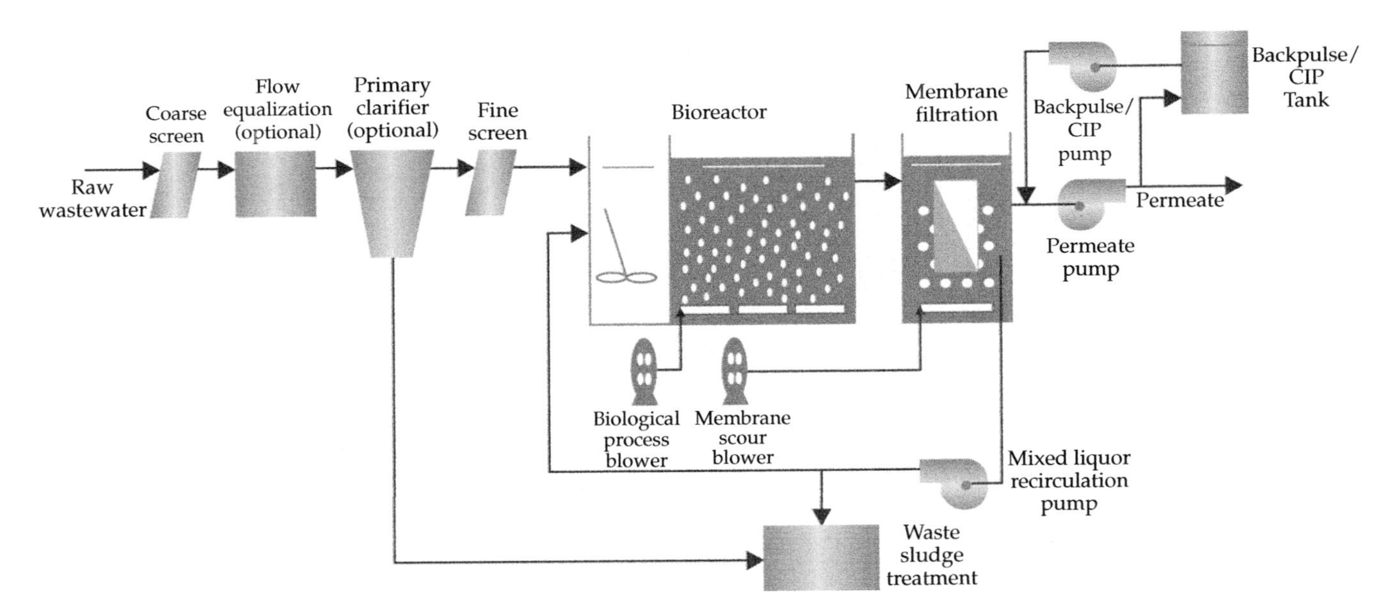

FIGURE 2.1 Example process flow diagram of a submerged MBR facility.

Although the effluent quality from an MBR system is dependent on the upstream biological process, MBRs are reliable at achieving an effluent with low solids (turbidity < 0.2 and total suspended solids < 1). Chapter 3 provides more detail on the effluent water quality that is achievable from an MBR system. Table 2.1 provides examples of some commercially available MBR equipment with general descriptions of membrane type and materials of construction. Design details for these products are provided in subsequent chapters.

1.2 Membrane Performance Metrics

There are several key terms that are important in membrane bioreactor system design and operations. Some of those key terms are defined here.

Although *flux* is used as the key design parameter for MBR systems, it is often the most misunderstood term in MBR vocabulary. By itself, *flux* is really a velocity term describing the volumetric filtration rate for a given area of membrane. A typical unit of flux is liters (gallons) (of water) per square meter (square feet) of membrane area per

TABLE 2.1 Examples of some commercial membrane manufacturers for MBR systems.

Manufacturer	Type	Material	Nominal pore size (μm)	Rating
GE	Submersible hollow fiber	Supported PVDF	0.04	UF[a]
Siemens	Submersible hollow fiber	Unsupported PVDF	0.04	UF
Kubota	Submersible flat sheet	Unsupported Chlorinated PE Polymer	0.4	MF[b]
Mitsubishi	Submersible hollow fiber	Unsupported Polyethylene Supported PVDF	0.4 0.4	MF
Toray	Submersible flat sheet	Unsupported PVDF	0.08	MF
Koch	Submersible hollow fiber	Supported Polyethersulfone	0.05	UF
Dow	Submersible hollow fiber	Unsupported PVDF	0.1	MF
Norit	Tubular	Supported PVDF	0.03	UF
Huber	Submersible rotating flat sheet	Unsupported Polyethersulfone	0.04	UF
Pall	Submersible follow fiber	Unsupported PVDF	0.1	MF

[a]UF = ultrafiltration.

[b]MF = microfiltration.

hour (day), or LMH *(gfd)*. Instantaneous flux is a measure of the amount of water filtered through a collection of membranes per unit of surface area for the outside membrane surface at a given moment. In practical terms, instantaneous flux is calculated simply by dividing an observed flowrate by the membrane area in service. *Average flux*, or *gross flux*, refers to the amount of water produced per day per membrane surface area in service during the filtration step. It does not account for downtime caused by physical (e.g., relaxation and backpulsing) or chemical cleaning requirements. *Net flux*, also referred to as *net production flux*, pertains to actual capacity or production capabilities. Net flux is equal to gross flux minus any production time lost and actual volume of permeate lost to relaxation and backpulsing or maintenance cleaning. Net flux does not take into account regeneration and recovery chemical cleanings. Examples for calculating instantaneous flux and net flux are provided in Appendix A. Typical design fluxes based on a survey of current MBR practices in significant design firms are presented in Chapter 5 (see Section 4.6, "Membrane Flux").

Transmembrane pressure (TMP) refers to the pressure differential across the membrane or the driving force required to achieve a given flux. Transmembrane pressure is measured by the difference in feed pressure and permeate pressure and should be adjusted for losses from the discharge to the pressure gauge. Membrane flux in immersed MBR systems is driven by either gravity head or a vacuum pumping system on the permeate side. In either case, it is difficult to measure feed-side and permeate-side pressure and estimates or averages are required. The unit of TMP is kPa (psi). Typical TMP operating ranges for different membrane types are presented in Chapter 4 (see Section 6.5, "Permeation").

Permeability is the instantaneous flux divided by TMP, usually expressed in LMH/kPa (gfd/psi). It is the amount of water per square meter (square feet) of membrane per unit time that can be pushed with 1 kPa (1 psi) of pressure.

2.0 MEMBRANE EQUIPMENT AND CONSTRUCTION MATERIALS

2.1 Membrane Classifications

Membrane use in wastewater treatment can generally be described as the use of a barrier to remove suspended, colloidal, or dissolved particles from the water source being treated. Membranes can be classified in the following five categories:

- Microfiltration,
- Ultrafiltration,

- Nanofiltration,
- Reverse osmosis, and
- Electrodialysis and electrodialysis reversal (ED/EDR).

As discussed in Chapter 1, microfiltration and ultrafiltration membranes are used in membrane bioreactors. Microfiltration and ultrafiltration membranes operate on size-exclusion mechanisms, unlike nanofiltration, reverse osmosis, and ED/EDR processes, in which other mechanisms are involved. For background purposes, a brief summary of the other membrane categories (i.e., nanofiltration, reverse osmosis, and ED/EDR) is also provided here.

Membrane classification is defined by pore size. Figure 2.2 illustrates the range of pore size for various membrane classifications and the relative size of various particles of interest. Membrane classification is more generally divided into those that remove dissolved particles (solutes) and those that remove suspended or colloidal particulates.

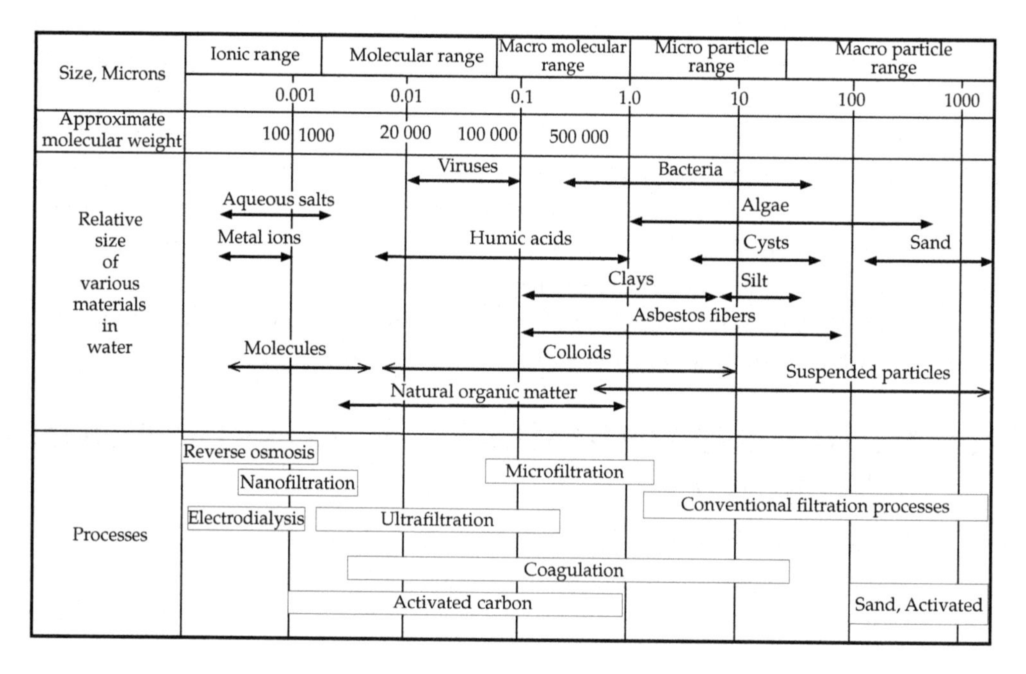

FIGURE 2.2 Membrane size-exclusion spectrum.

Membranes that remove solutes include reverse osmosis, nanofiltration, and ED/EDR. Reverse osmosis and ED/EDR remove the smallest ionic particles. Reverse osmosis is capable of removing ionic species and synthetic organic compounds by diffusion and sieving through a semi-permeable membrane. Electrodialysis and electrodialysis reversal are also capable of removing ionic species through the use of cation- and anion-selective membranes; however, ED/EDR does not remove pathogen or organic compounds. Nanofiltration operates the same as reverse osmosis; however, nanofiltration has a slightly larger pore size than reverse osmosis. Nanofiltration is effective at removing multivalent ions and synthetic organic compounds, although it is less effective at removing monovalent ions. Nanofiltration is often referred to as *membrane softening* because it is capable of removing multivalent ions such as calcium and magnesium.

Both the reverse osmosis and nanofiltration processes use the reversal of natural osmotic pressure by applying a pressure that is greater than the osmotic pressure to the concentrated solution. The pressure required is highly dependent on the amount of dissolved solids in the feed water, and can range from less than 700 kPa (100 psi) on relatively freshwater nanofiltration applications to more than 7000 kPa (1000 psi) for reverse osmosis applications on seawater (Allgeier et al., 2005). The driving force for ED/EDR is an electrical charge applied to the membranes allowing salt to move through the membrane using electrical potential. In the electrodialysis reversal process, the direction of ion flow is periodically reversed by reversing the polarity-applied electric current to reduce accumulated ions from the membrane surface to minimize the effects of scaling and fouling.

Because reverse osmosis and nanofiltration operate by diffusion through semi-permeable membranes rather than size exclusion, they are characterized by their molecular weight cutoff (MWCO) rather than a defined pore size. The MWCO for nanofiltration typically ranges from 200 to 1000 daltons, whereas the typical MWCO for reverse osmosis membranes is less than 100 daltons (Allgeier et al., 2005). Reverse osmosis and nanofiltration membranes are also characterized by the percent of salt rejection that can be achieved through the membrane process. Salt rejection is typically defined by the percentage of monovalent ion rejection for reverse osmosis membranes and by monovalent and divalent ions for nanofiltration membranes.

Membranes that remove suspended or colloidal particles include microfiltration and ultrafiltration. Microfiltration and ultrafiltration operate through an actual physical barrier between the membrane and the particle to be removed. For this reason, microfiltration and ultrafiltration are often referred to as *membrane filtration*. Microfiltration and ultrafiltration membranes have a distribution of pore sizes that may vary depending on the material and the manufacturer, although microfiltration membranes typically have pore sizes in the range of 0.1 to 0.4 μm and ultrafiltration membranes typically have pore sizes in the range of 0.01 to 0.05 μm. The pore sizes of both microfiltration and ultrafiltration membranes are discussed in the following sections. Both microfiltration and ultrafiltration are effective at removing fine particles and providing a barrier to bacteria, *Cryptosporidium*, *Giardia*, and some viruses (ultrafiltration, in particular).

Both microfiltration and ultrafiltration operate through either a pressure drive or a vacuum-driven process in which the clean water (filtrate) is "pushed" (pressure-driven) or "pulled" (vacuum-driven) through the membrane, which creates TMP. Transmembrane pressure requirements for microfiltration and ultrafiltration membranes are less than the pressure feed requirements for nanofiltration and reverse osmosis membranes. Transmembrane pressure can vary based on membrane properties, number and size of particles, frequency of backwashing and cleaning, flux, and the fouling potential of the source water. Transmembrane pressure typically ranges from 20 to 100 kPa (3 to 14.5 psi) (Wilf et al., 2010).

Microfiltration and ultrafiltration membranes are typically associated with MBRs because of their removal mechanism (i.e., physical barrier for particle removal). Nanofiltration and reverse osmosis are typically used in wastewater treatment as a purification step for treated wastewater effluent that requires the removal of trace organics, trace inorganics, and microbes prior to discharge. Therefore, a nanofiltration or reverse osmosis process at a wastewater treatment facility could be located downstream of an MBR process.

2.1.1 *Microfiltration*

The pore size for microfiltration membranes has traditionally been defined as 0.1 to 0.2 μm (nominally, 0.1 μm). However, in some MBR applications, the pore size is larger than that used for traditional membrane filtration applications, with pore sizes as large as 0.4 μm (Wilf et al., 2010). Larger pore sizes are used for certain industrial applications and can also be used to reduce the potential for fouling. The microfiltration membrane pore size is effective at providing a physical barrier for particulates, bacteria, and protozoan cysts, but not viruses. Microfiltration membranes do provide some removal of viruses caused by the buildup of particulates on the membrane surface; however, they do not provide an effective barrier for viruses.

2.1.2 *Ultrafiltration*

The pore size for ultrafiltration membranes is an order of magnitude finer than microfiltration membranes and is typically defined as 0.01 to 0.05 μm (nominally, 0.01 μm). Ultrafiltration membranes are also capable of removing larger organic macromolecules and, therefore, have also been defined by a MWCO (Allgeier et al., 2005). Typical MWCO levels for ultrafiltration membranes range from 10 000 to 500 000 daltons. The ultrafiltration membrane pore size is effective at providing a physical barrier to particulates, bacteria, protozoan cysts, and some viruses.

2.2 Membrane Materials

Membrane material selection entails both manufacturing process and performance considerations. The most important noncost factor in the manufacturing process is the ability to produce a well-defined pore-size rating. Performance considerations include strength and integrity, resistance to fouling, resistance to chemicals, membrane life, peel strength (for supported membranes), resistance to temperature, bubble-point test pressure, and shrinkage over time. The majority of membranes are manufactured from polymeric materials. Table 2.2 provides a summary of the characteristics of these membrane materials.

Ceramic membranes are now available for ultrafiltration and microfiltration applications. Historically, ceramic membranes have been considered too expensive and, thus, have not had a significant effect on the municipal wastewater market. However, because of successful demonstrations of their high integrity, chemical resistance, and low operating costs, there have been recent market advances of ceramic membranes (Wilf et al., 2010). Because integrity is not as critical in wastewater treatment applications, ceramics are not typically used in these applications (Wilf et al., 2010).

Cellulose acetate is a polymeric material that is fully hydrophilic (i.e., having an affinity for water or readily absorbing or dissolving in water). Although cellulose acetate membranes have been used for nanofiltration and reverse osmosis applications, they are not used in MBR applications as they are not suitable for wastewater feeds because of their biodegradability (Wilf et al., 2010). Cellulose acetate membranes are also low in strength and elongation.

Table 2.2 Summary of membrane material characteristics (Wilf et al., 2010).

Material	Characteristics	Advantages	Disadvantages
Polyethersulfone (PES)	Moderately hydrophilic, high strength, low elongation	Strength, high permeability	Limited pressure application for air scour, lower chlorine tolerance
Polysulfone	Moderately hydrophilic with additives, high strength, low elongation	Strength, permeability	Limited pressure application for air scour, lower chlorine tolerance
Polyvinylidene fluoride (PVDF)	Relatively hydrophobic, high strength, high elongation	Chemically resistant, strength, high chlorine resistance, long membrane life, suitable for use with air scour, permeability	More expensive, slightly lower permeability than PES
Polypropylene	Hydrophobic	Chemically resistant, resistant to biodegradability	Susceptible to oxidation
Polyethylene	Hydrophobic	Chemically resistant, resistant to biodegradability	Susceptible to oxidation

2.2.1 *Polymeric*

Polymers are the most common materials used in membrane manufacturing. There are several types of polymeric membranes available on the market today, including polyacrylonitrile, polyethersulfone (PES), polysulfone, polyvinylidene fluoride (PVDF), polypropylene, and polyethylene. The main distinguishing factors in polymer selection are hydrophilic and hydrophobic properties, strength, and elongation.

Polyacrylonitrile, PES, polysulfone, and PVDF are in between fully hydrophilic and fully hydrophobic. Each of these polymers is naturally hydrophobic, but can be blended with additives and pore formers to make a moderately hydrophilic membrane

(Wilf et al., 2010). Some of the advantages and disadvantages of these polymers were outlined in Table 2.2. The two most common membrane materials in MBR applications are PES and PVDF (Wilf et al., 2010). Both PES and PVDF have good strength, whereas PVDF has a much higher elongation. Strength and elongation are important considerations in terms of susceptibility for fiber breaks. Polyvinylidene fluoride and PES provide good resistance to chemicals; however, PES has a lower chlorine tolerance than PVDF.

Polyethylene and polypropylene are both fully hydrophobic polymers. While they provide good resistance to chemicals and biodegradability, they are difficult to wet and susceptible to drying out during air-based integrity tests (Wilf et al., 2010). Fully hydrophobic polymers are not as common in membrane applications as the polymers that can be modified to be moderately hydrophilic.

Polytetrafluoroethylene (PTFE) is an emerging polymeric material in MBRs; it exhibits good strength and chemical and temperature resistance. Although PTFE is not as widely used as some of the other membrane materials, it is beginning to be used and marketed in MBR applications.

2.2.2 *Composite*

Composite membranes are engineered materials made from two or more materials, one of which constitutes the active surface while another is considered the supporting layer. The active layer is typically a thin dense layer that serves as the filtration barrier (Allgeier et al., 2005). Composite membranes increase productivity by combining high-rejection properties of the active layer with optimum porosity of the support layer (Allgeier et al., 2005). The support structure is more porous to improve flow through the membrane. Composite membranes are typically used for nanofiltration and reverse osmosis membranes, but not microfiltration or ultrafiltration membranes. Therefore, they are not used in MBR applications.

3.0 MEMBRANE SYSTEM CONFIGURATIONS

3.1 Membrane and Membrane-Element Types

Three main membrane-element types are discussed in this section based on the type of membrane used (i.e., hollow-fiber, flat-sheet, and tubular membrane-element types). Table 2.3 presents a summary of the membrane types, including their advantages and disadvantages (Li et al., 2008; Lesjean et al., 2004).

TABLE 2.3 Summary of advantages and disadvantages of membrane types (based on Li et al. [2008] and Lesjean et al. [2004]).

Criteria	Hollow fiber	Flat sheet	Tubular
Packing density	High	Moderate	Low
Operational flux	Moderate	Moderate to high	High
Clogging propensity	Moderate	Moderate	Low
Turn up/down	Limited by TMP	Limited by TMP	Good
Cleaning	Moderate (backflush possible)	Poor (no backflush)	Good
Energy use	Moderate	Moderate to high	High
Capital cost	Moderate	Moderate to high	High

3.1.1 *Hollow Fiber*

Hollow-fiber membrane-element types are the most common configuration used in MBR technologies, with around 25 commercial products available. The fibers have an outside diameter varying from 0.5 to 3 mm. An element is formed by sealing the hollow fibers to one or two headers (at opposite sides), which can be circular or rectangular. The individual hollow fibers are potted as one dense bundle, several subbundles, or are spaced out individually at the location of the header. The hollow-fiber membrane elements are submerged in the biomass without a pressure vessel. As a result, the biomass feed is located outside of the fiber. Pumping wastewater with high suspended solids inside the bore of the hollow-fiber membrane would lead to severe plugging because of the narrow inner diameter of hollow fibers. These elements are grouped together in a frame or rack, which are operated as one unit. The fiber orientation in a rack is mainly vertical, with some manufactures offering a horizontal orientation. Figure 2.3 presents a typical hollow-fiber membrane configuration based on Li et al. (2008). The retentate is retained in the membrane tank and exits the system through sludge wasting and draining of the tank during backwashing and cleaning.

Each membrane configuration has inherent advantages and disadvantages. The hollow fiber membrane configuration are mostly made out of ultrafiltration membranes (where most flat sheet membranes are made out of microfiltration membranes) and are characterized by a high-fiber packing density. As a result, hollow-fiber membrane MBR systems are operated at a moderate operational flux, but with a high

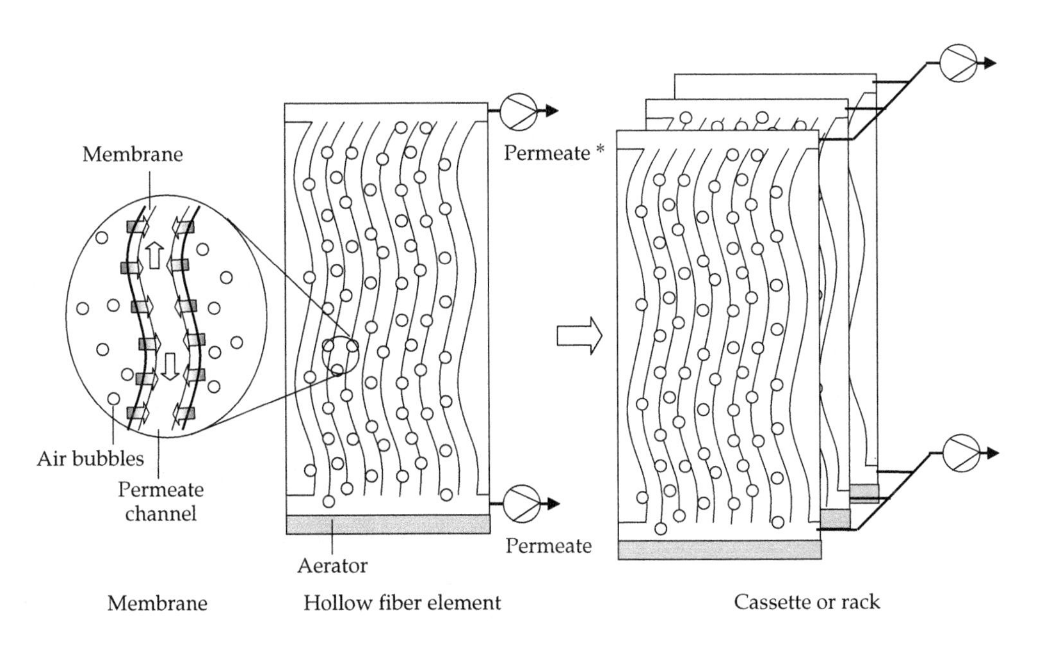

FIGURE 2.3 Membrane-element type—hollow fiber (based on Cornel and Krause [2008]) (*applicable for submerged hollow-fiber systems with permeate extraction from both sides).

propensity to clog the membrane channels. These hollow-fiber MBR systems are immersed and, as a result, are limited by TMP to turn up the capacity. Cleaning is done by chemical and physical means, including the possibility to backwash. Energy use and capital costs are moderate and are mainly based on air scouring and membrane cost, respectively.

3.1.2 *Flat Sheets and Plates*

Flat-sheet membranes and plates are membrane envelopes that have two membrane sheets connected by an internal support structure that serves as the permeate extraction channel. This structure is called a *membrane panel*; most flat-sheet panels or plates are rectangular. The overall thickness of the membrane panel varies between 5 and 13 mm. In most instances, these membrane panels are installed vertically in a membrane cassette, with a specific spacing in between each membrane sheet between 6 and 10 mm. In some instances, rotating plate modules are used. Figure 2.4 presents a typical configuration for flat-sheet membranes and plates based on Li et al. (2008).

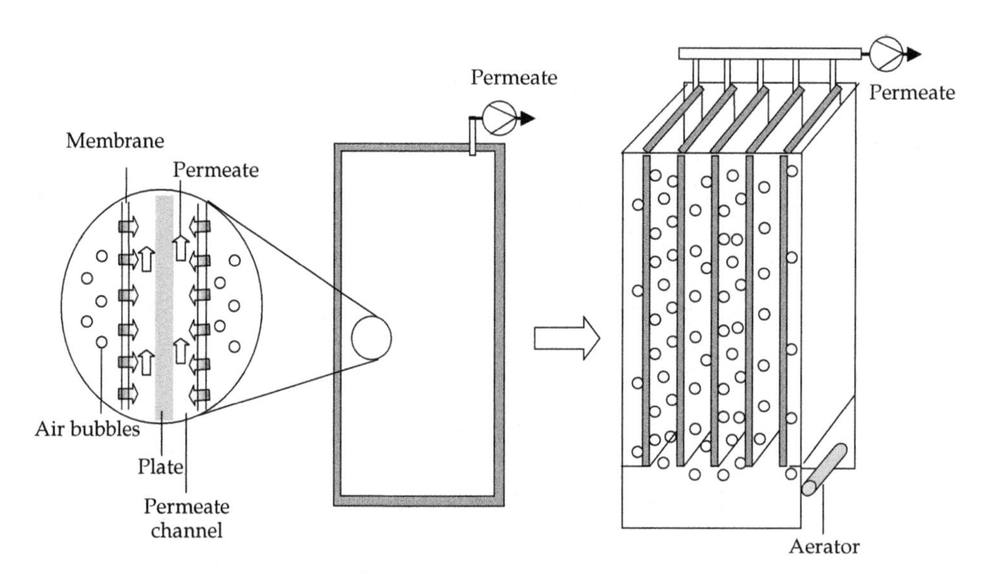

FIGURE 2.4 Membrane-element type—flat-sheet plates (based on Cornel and Krause [2008]).

The retentate is retained in the membrane tank and exits the system through sludge wasting and draining of the tank during backwashing and cleaning.

Flat-sheet membranes are typically microfiltration membranes with a high permeability, which enables operation at a moderate-to-high operational flux. Flat-sheet membrane configurations have a moderate packing density and a moderate propensity for clogging of the membrane channels. These flat-sheet MBR systems are immersed and, as a result, are limited by TMP to turn up the capacity. Cleaning efficiency depends primarily on chemical cleaning and air scouring efficiency only, while there is no backwash or other means of shear promoters. As a result, the energy consumption and capital cost of flat sheets is moderate to high.

3.1.3 Tubes

Tubular membranes are large-diameter tubes with an inner diameter ranging from 5 to 13 mm. These membranes are supported by a mechanically stable polyester backing. Tubes are sealed on both ends in a membrane module and are surrounded by a plastic shell. As such, the biomass feed is located at the inside of the tube where the membrane skin (i.e., discriminating layer) is located. They are installed vertically as individual modules or horizontally as a series of modules in one pressure vessel.

Tubular membranes have a low packing density, which originates from their internal diameter. Tubular membranes are contained in a pressure vessel and are installed as a sidestream application (i.e., the membranes are mounted externally from the biological process tank), which allows for freedom to turn the treatment capacity up or down. Furthermore, these membranes can be operated at a higher operational flux and at higher mixed liquor suspended solids (MLSS) concentrations. This external MBR configuration uses no air scouring (except for the vertical system with air lift) and has limited cleaning requirements. Clean-in-place procedures are simplified in sidestream tubular MBR systems (e.g., does not require the draining of the biological process tank) because of the containment of the permeate and retentate streams. In addition, cleaning efficiency is good because of a high-shear velocity under cross-flow filtration (with or without air) and some ability to backflush. Energy cost is high because of the high pumping rate for cross-flow filtration. Figure 2.5 presents a typical configuration for tubular membranes based on literature by Li et al. (2008).

3.1.4 Membrane Component Operation (Membrane Process Configurations)

Two main process configurations that are used today are internal MBR ("iMBR," or immersed) and external MBR ("eMBR," or sidestream). The sidestream MBR process has membranes mounted externally (comparable to pressurized ultrafiltration). Figure 2.6 presents these two membrane configurations.

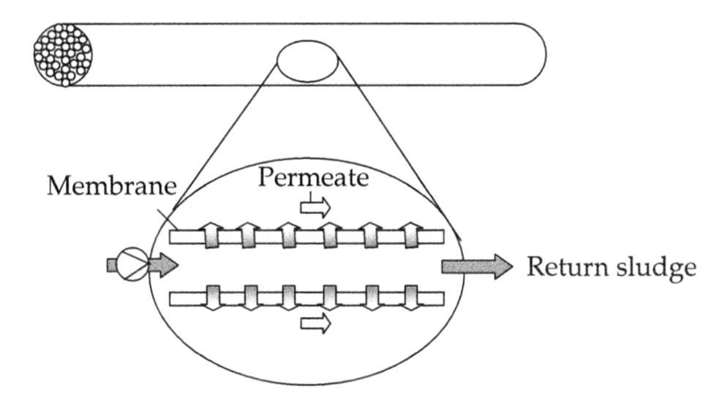

FIGURE 2.5 Membrane-element type—tubular membranes (based on Cornel and Krause [2008]).

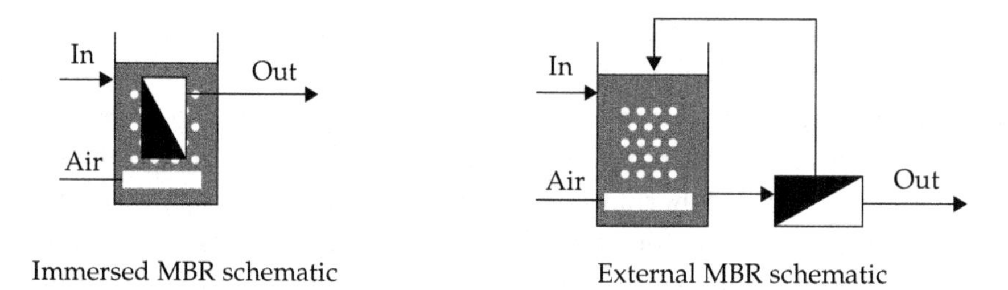

FIGURE 2.6 Membrane component operation.

3.1.4.1 Internal (Immersed) An internal MBR has membrane elements grouped in a frame or rack immersed in the biological tank of a wastewater treatment system (Figure 2.6). In this configuration, the biologically active sludge is located at the outside of the fibers. As a result, these membranes are operated under suction and the permeate is flowing from the outside of the membrane to the inside of the membrane (i.e., "outside-in"). Coarse-bubble aeration is provided to promote shear at the outside of the membrane surface. This process configuration is used for hollow-fiber and flat-sheet membrane types.

3.1.4.2 External (Sidestream) External MBR systems have the membrane unit operation separated from the biological tank of a wastewater treatment system. The sludge is pumped into the membrane modules.

Membrane types with a discriminating layer at the outside are operated under suction and the permeate flows from the outside of the membrane to the inside. This process configuration is used for tubular membrane types.

Membrane types with a discriminating layer at the inside are operated under pressure and the permeate flows from the inside of the membrane to the outside. This process configuration is used for tubular membranes.

3.1.4.3 Constant Flux Versus Constant Pressure Operation Membrane bioreactors can be designed to operate at either a constant flux or a constant pressure (i.e., TMP). The current trend in MBR design is to operate at constant flux (Le-Clech et al., 2006). For constant flux operation, the TMP is operationally adjusted as required to maintain a constant flux across the membrane in response to changes in permeability that occur during operation as a result of fouling. For constant pressure operation, a constant TMP is maintained across the membrane and the flux will vary in response to the changes in permeability. Further discussion of fouling and its impact on MBRs is discussed in the following section, with additional information presented in Chapters 4 and 7.

4.0 MEMBRANE OPERATION AND MAINTENANCE FUNDAMENTALS

4.1 Membrane Fouling

The most important O&M consideration for MBRs is membrane fouling, which directly impacts the performance of the MBR process. Failure to adequately control or reduce membrane fouling results in substantially higher O&M and replacement costs for the MBR process and, in worst cases, may result in an inability to treat the required design flows.

4.1.1 *Effect on Membrane Performance*

Membrane fouling occurs when material deposits and accumulates on the membrane surface or in the membrane pores, which reduces the permeability of the membrane. In an MBR, this material typically consists of suspended particulates (primarily microorganisms and cell debris), colloids, and solutes (Meng et al., 2009).

Depending on the mode of operation, this impacts membrane performance in one of two ways. For constant pressure operation, fouling results in a reduction of permeate flux. In comparison, during constant flux operation, membrane fouling results in an increase of TMP.

4.1.2 *Types of Membrane Fouling*

Based on applicable treatment and cleaning options, membrane fouling can be characterized as physically reversible, chemically reversible, or irreversible. Physically reversible fouling can be removed using some form of physical cleaning or treatment, and is caused by loosely attached foulants that are the easiest to clean and remove. Chemically reversible fouling requires chemical cleaning to remove foulants, which are more strongly attached to the membrane surface or inside pores and, therefore, are more difficult to clean and remove. Irreversible fouling is permanent and cannot be cleaned by either physical or chemical means. Other terms used for irreversible fouling may include untreatable fouling, irrecoverable fouling, permanent fouling, or absolute fouling.

In the past, there has been some confusion associated with the term *irreversible fouling* because of alternative definitions used by other publications. However, in this MOP, use of the term *irreversible fouling* is limited to fouling that is permanent and cannot be cleaned by either physical or chemical means.

4.1.3 *Membrane Fouling Mechanisms*

Membrane fouling occurs through a variety of mechanisms. *Cake fouling* refers to the physical accumulation of colloidal and suspended material above the membrane surface. This material is larger in size than the membrane pores and forms a cake layer at the surface, which offers an additional resistance for filtration (Le-Clech et al., 2006). Cake fouling is physically reversible, and is graphically presented in Figure 2.7.

Pore blocking, which is graphically presented in Figure 2.8, occurs when colloids, solutes, and microbial cells precipitate inside the membrane pores, and may include the precipitation of inorganic compounds. Pore blocking is typically chemically

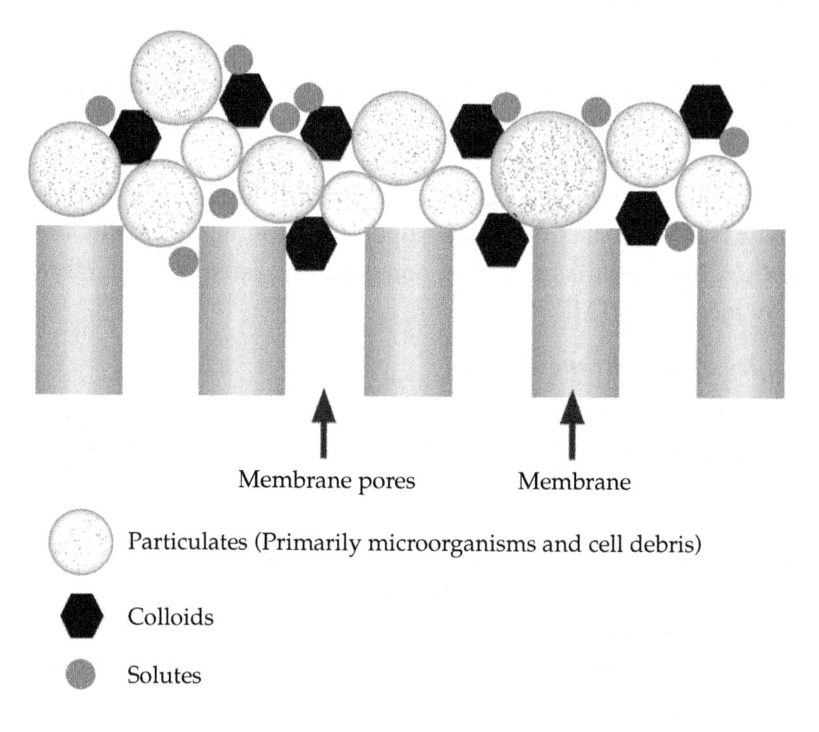

FIGURE 2.7 Membrane fouling mechanisms—cake fouling.

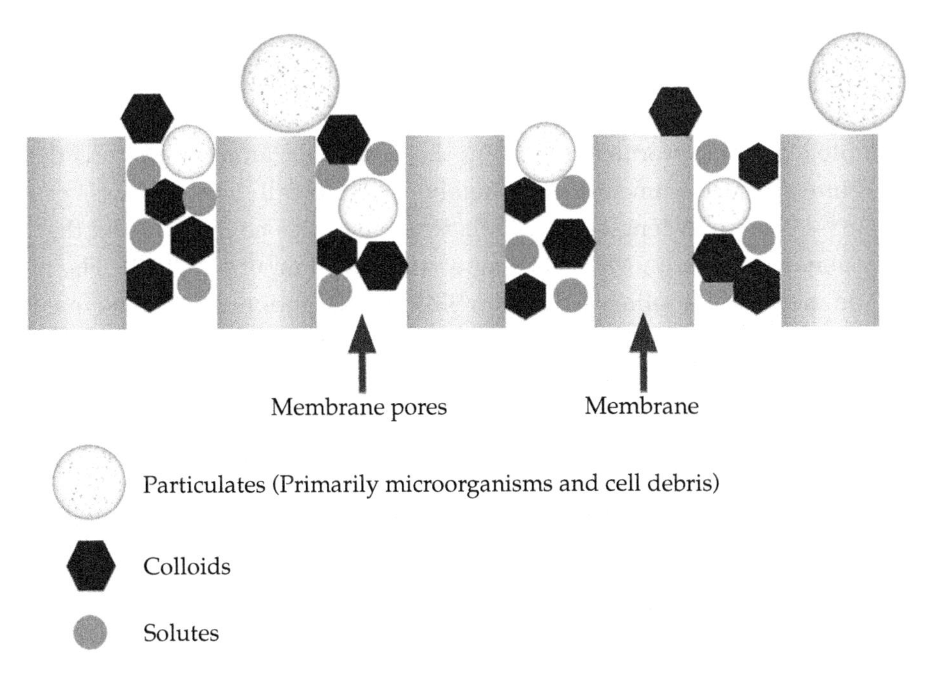

Figure 2.8 Membrane fouling mechanisms—pore blocking.

reversible, but, in some instances, may be irreversible, particularly for certain types of inorganic fouling (Meng et al., 2009).

Research has indicated that extracellular polymeric substances (EPS) and soluble microbial products (SMP) are key foulants in MBRs (Pan et al., 2010). *Extracellular polymeric substances* and *SMP* both refer to a variety of macromolecules such as polysaccharides, proteins, nucleic acids, phospholipids, and other polymeric compounds that are produced by active secretion, shedding of cell surface material, or cell lysis and are used for the formation of flocs and biofilms (Le-Clech et al., 2006).

Extracellular polymeric substances represent a solid matrix that is bound or floc-associated, whereas SMP are soluble, as the name implies. Extracellular polymeric substances have been linked to fouling and cake resistance by several studies, with higher concentrations of EPS resulting in increased cake resistance and higher fouling rates. Concentrations of EPS are also closely connected to activated sludge characteristics that affect membrane performance and fouling, including flocculation ability, hydrophobicity, surface charge, and sludge viscosity. Consequently, EPS play an important role in membrane performance and fouling in MBRs (Meng et al., 2009).

Soluble microbial products represent the pool of organic compounds that are released into solution from substrate metabolism (typically with biomass growth) and biomass growth. (Meng et al., 2009). Recent studies have indicated that SMP are also major foulants in MBRs (Meng et al., 2009; Pan et al., 2010). The effects of SMP on membrane fouling depend on SMP concentration, membrane materials, and operation modes, with some studies indicating that polysaccharide-like substances in SMP contribute to fouling more than protein-like substances. Therefore, control of SMP concentrations in MBRs is an important factor in fouling control, which can be achieved by controlling operating parameters such as solids retention time (SRT), hydraulic retention time (HRT), food-to-microorganism (F/M) ratios, and dissolved oxygen concentrations. Most of the reported results from recent studies indicate that the SMP concentration decreases with increasing SRT, with higher concentrations of SMP at lower dissolved oxygen concentrations (Meng et al., 2009).

4.1.4 *Critical Flux*

The concept of critical flux has been used to describe the relationship between flux and the fouling rate in controlled steady-state environments (Wilf et al., 2010). There are several definitions of *critical flux* in the literature, including "that flux below which no particulate fouling occurs" (Wilf et al., 2010) and "the flux rate above which rapid fouling and reduction of permeability occur" (Judd, 2006). Use of the critical flux concept in practical applications can be problematic because MBRs are not operated under steady-state conditions and the fouling rate is a highly variable parameter. Therefore, because there is some confusion associated with the term *critical flux*, the concept is not recommended for evaluating MBR technologies or applications. Rather, the term *sustainable flux*, as defined in the following section, is recommended for use by the wastewater practitioner.

4.1.5 *Sustainable Flux*

Sustainable flux is the flux at which a modest degree of fouling occurs that can be handled by the MBR's fouling control mechanisms, providing an acceptable compromise between capital and O&M costs for MBRs (Wilf et al., 2010). When an MBR is operated beyond the sustainable flux, the rate of fouling will exceed the capability of the control mechanisms to remove foulants and membrane permeability will decrease at a rate that cannot be sustained operationally. Sustainable flux is site-specific and is a function of wastewater characteristics, the operating parameters listed previously (SRT, HRT, and the F/M ratio), and the cleaning methods and frequency used. As referenced previously, typical design fluxes are presented in Chapter 5 based on a survey of current MBR practices in larger design firms (see Section 4.6, "Membrane Flux").

4.2 Biological Factors

Biological process conditions represent one of the most important factors affecting membrane fouling and performance in MBRs. These factors are complex and varied, and include SRT, HRT, MLSS concentrations, sludge viscosity, sludge hydrophobicity, EPS and SMP production, and filamentous bulking.

Solids retention time, which is one of the primary biological process operation and control parameters, is also one of the most important parameters affecting membrane fouling in MBRs. Studies have shown that operation at extremely low SRTs (down to 2 days) substantially increases membrane fouling (Le-Clech et al., 2006), which has been attributed to increased concentrations of SMP and EPS at lower SRTs (Meng et al., 2009). Similarly, operation at SRTs greater than 50 days has also been shown to increase membrane fouling and reduce membrane performance based on a combination of factors, including increased MLSS concentrations and high sludge viscosity (Meng et al., 2009). These competing factors must be balanced to achieve an optimal SRT operating range (and corresponding MLSS concentration range) for a particular wastewater and membrane configuration. Appropriate SRT design ranges for MBRs are discussed in more detail in Chapter 4.

Recent research has also indicated that filamentous bulking can substantially increase membrane fouling, primarily through increased production of SMP (Pan et al., 2010). Other factors associated with filamentous bulking that have been linked to increased membrane fouling include increased sludge viscosity and sludge hydrophobicity and increased concentrations of bound EPS (Meng et al., 2009). It is important to note that although the membranes will continue to produce a high-quality effluent even with filamentous bulking, increased membrane fouling caused by filamentous bulking is a critical design consideration that affects sustainable flux. Filamentous bulking can be controlled through a variety of design parameters and operational techniques, including SRT and F/M ratio control, dissolved oxygen control, use of selectors, selective wasting, foam control, and chemical addition

(WEF et al., 2009), which are discussed in more detail in later chapters. Hydraulic retention time has also been linked to membrane fouling in several studies, with reductions in HRT resulting in increased membrane fouling because of increases in sludge viscosity and EPS concentrations (Meng et al., 2009).

4.3 Fouling Control

4.3.1 *Introduction*

There are a variety of physical and/or chemical methods in MBRs that are used to reduce or control fouling and to maintain or restore permeability. These methods are described in the following section and are also discussed in more detail in Chapter 4 (see Section 6.6, "Fouling Control") and Chapter 7 (see Section 2.1, "Membrane Cleaning").

4.3.2 *Air Scour*

Air scour is a physical cleaning method used to prevent or reduce membrane fouling in submerged MBR configurations. Aeration is used to generate a shear stress at the membrane surface, which provides a scouring action and reduces membrane cake fouling. However, excessive aeration can create stress on the membrane, potentially resulting in breakage or fatigue, and reduce the overall membrane life. Excessive aeration can also lead to the breakage of sludge flocs and the production of SMP, increasing the membrane fouling potential, so there is an optimal range of air scour rates (Meng et al., 2009). In addition, the energy requirements associated with continuous air scour are significant. Intermittent, cyclic, or varying air scour flowrates can be used to reduce the energy requirements associated with air scour and, under certain conditions, may result in better fouling control.

If there is insufficient air scouring, localized dewatering may occur when concentrated solids accumulate at the membrane surface. There are numerous terms used in the industry to describe the cause and effect of localized dewatering, including *clogging*, *sludging*, and *plugging*.

4.3.3 *Relaxation*

Relaxation involves the temporary cessation, or pausing, of membrane production while maintaining air scour. Membrane relaxation can improve membrane production by enhancing the back transport of foulants away from the membrane surface through the concentration gradient, which is further enhanced by maintaining air scour (Le-Clech et al., 2006).

4.3.4 *Adsorbent and Coagulant Addition*

The addition of adsorbents and coagulants to mixed liquor can also be used to reduce fouling in MBRs. Ferric chloride, alum, and powdered activated carbon have all shown the ability to reduce SMP concentrations when added at the appropriate dosages, thereby reducing fouling and improving membrane performance (Le-Clech et al., 2006). Similar coagulants, such as poly aluminum chloride, are expected to provide comparable benefits. Recently, some modified cationic polymers have been introduced for use in MBRs as membrane performance enhancers; these operate on the same principle of reducing SMP concentrations (Le-Clech et al., 2006). Potential advantages include being able to operate the membranes at a higher sustainable flux or reduced air scour, which can result in energy savings. Disadvantages include the O&M costs associated with additional chemical feeds.

It is important to note that some of the aforementioned chemicals (e.g., ferric chloride, poly aluminum chloride, and alum) are also commonly used for phosphorus precipitation and removal for advanced wastewater treatment (WEF et al., 2009). Therefore, if these chemicals are applied to an MBR process for phosphorus removal, an additional operational benefit could be reduced fouling and improved membrane performance. In comparison, use of lime for chemically enhanced primary treatment or phosphorus precipitation and removal upstream of an MBR could have adverse effects on membrane performance because of additional inorganic fouling that may result from the addition of calcium and the pH adjustment associated with lime; studies have shown that the carbonates of calcium and magnesium can increase the potential of membrane scaling. Metal ions may also play a significant role in the formation of fouling layers, which can bridge the deposited cells and biopolymers and form a dense cake layer (Meng et al., 2009). The actual effects on membrane fouling will be site-specific depending on the water chemistry involved; as such, the wastewater practitioner will need to consider potential effects on membrane fouling when selecting treatment chemicals and dosages for use in MBR applications. For additional information regarding typical chemical dosages used for advanced wastewater treatment, refer to Chapter 16 ("Physical and Chemical Processes for Advanced Wastewater Treatment") of *Design of Municipal Wastewater Treatment Plants* (WEF et al., 2009).

4.3.5 *Backwash Cleaning*

Backwash cleaning refers to any instance where water (typically permeate) is charged to the membranes in the reverse direction of permeate flow and may include chemical addition to enhance cleaning and fouling control. Other commonly used terms include *backflushing* and *backpulsing*. Backwash cleaning, which is of short duration, is performed in situ and is primarily used to remove physical fouling.

4.3.6 Chemical Cleaning

To address chemically reversible fouling, cleaning with chemical solutions of sodium hypochlorite for removal of organic foulants, citric acid or oxalic acid for removal of inorganic foulants, other mineral acids, or caustic are required. There are a number of definitions currently used for various chemical cleaning techniques. These fall into one of the following two categories:

- *Maintenance cleaning*, which involves charging cleaning solution to the membranes in the reverse direction of permeate flow for a short duration (typically between 0.5 and 1.0 hour). Depending on the application, the actual cleaning time can last up to 6 hours or longer. The procedure is performed in situ with the mixed liquor or activated sludge remaining in the tank. Another commonly used term for maintenance cleaning is *chemically enhanced backwash* or *regeneration cleaning*.
- *Recovery cleaning*, which can be performed in situ or ex situ, is conducted by charging the cleaning chemicals to the membranes in either direction. The procedure is performed in the absence of mixed liquor or activated sludge, and typically lasts longer than 2 hours. The actual duration is variable and is a function of the application and degree of fouling involved, the cleaning water temperature (warmer temperatures will result in a quicker clean), and the cleaning chemical concentration. If performed in situ, the tank is first drained of mixed liquor and then refilled with a dilute cleaning solution. Other commonly used terms for recovery cleaning include *intensive cleaning* and *CIP*. Recovery cleaning is typically a nonroutine procedure that is performed once or twice a year, with a maximum frequency of no more than once a month.

4.3.7 Mechanical Cleaning

In some instances, the cleaning options described previously are not sufficient and the membranes must be physically cleaned by hand or sprayed down with water, which is referred to as *mechanical cleaning*. This is typically required when rags, debris, or dewatered solids form a concentrated cake on the membranes that is difficult or impractical to remove by other means, and is typically associated with insufficient pretreatment or localized dewatering. Care must be taken to avoid using power sprays or high-pressure hoses or nozzles that could physically damage the membranes. As such, it is recommended that the membrane manufacturer be consulted regarding acceptable water pressure ranges and mechanical cleaning methods to avoid damaging the membranes. Other common terms for mechanical cleaning include *manual cleaning*, *hand cleaning*, or *physical cleaning*.

4.4 Membrane Failure Mechanisms

4.4.1 *Irreversible Fouling*

Irreversible fouling refers to fouling that is permanent and cannot be removed by any of the physical, mechanical, or chemical cleaning techniques described previously. Irreversible fouling is typically associated with pore blocking, which is more difficult to remove than cake fouling. Inorganic fouling, which may not be easily eliminated even by recovery cleaning, may also play a significant role in irreversible fouling (Meng et al., 2009).

4.4.2 *Membrane Integrity*

Much attention has been paid on the integrity of microfiltration and ultrafiltration membranes because of their use in drinking water applications as a barrier to *giardia*, *cryptosporidium*, and some viruses (ultrafiltration). Although membrane integrity does not carry the same weight in wastewater applications, it is important to process performance, membrane life, and maintenance.

Membrane integrity is defined by the integrity of the individual membrane elements. Failures in membrane integrity can be divided into two categories, manufacturing defects and operational problems. Manufacturing defects include the following (Wilf et al., 2010):

- Macroporous voids resulting from manufacturing defects. These voids can potentially be a source of weakness and can result in breakage.
- Breaks from mishandling in module manufacture or as a subsequent breach from a physical weakening during module assembly.
- Slits caused by membranes crushed during manufacturing, creating a potential for fatigue failure during normal pressure cycling operation.
- Problems with the epoxy potting procedures giving weakness near the tube sheet and causing membrane embrittlement.

Operational problems that lead to integrity failures include the following (Wilf et al., 2010):

- Slits caused by collapse from over-pressurization and the formation of a stress fracture caused by water hammer or excessive pressure during backwash,
- Scratches and piercing from debris during operation caused by screen failure or permeate system contamination, and
- Embrittlement and pore collapse from chemical attack.

The most obvious cause of integrity issues is allowing large particles to come in contact with the membrane surface through the pretreatment process. Adequately sized and operated screens are the best defense against allowing damaging particles through. Particles in the permeate side of the system can also damage the membrane. Care should be taken to ensure all pipes and tanks downstream of the membrane system are properly cleaned and flushed to prevent particles from making their way into the permeate size of the membrane during backwashing or cleaning. In addition, coatings and/or linings of tanks, pipes, and other equipment may flake off and damage the membranes; as such, extra care must be taken when selecting, specifying, applying, and maintaining coating and lining systems in a membrane system to prevent this from occurring. Selecting proper cleaning chemicals and dosages also helps prevent the weakening or embrittlement of membranes. Finally, proper control of pumps, air supply, valves, and equipment during backwashing and air scouring help prevent over-pressurization or water hammer during the backwash and air scouring modes. These operational considerations are also affected by proper design of the system to reduce the likelihood of integrity issues arising.

5.0 REFERENCES

Allgeier, S.; Alspach, B.; Vickers, J. (2005) *Draft Membrane Filtration Guidance Manual;* U.S. Environmental Protection Agency: Cincinnati, Ohio.

Cornel, P.; Krause, S. (2008) Membrane Bioreactor for Wastewater Treatment. In *Advanced Membrane Technology and Applications;* Wiley & Sons: Hoboken, New Jersey; pp 217–236.

Judd, S. (2006) *The MBR Book: Principles and Applications of Membrane Bioreactors in Water and Wastewater Treatment;* Elsevier: Oxford, U.K.

Le-Clech, P.; Chen, V.; Fane, T. A. G. (2006) Fouling in Membrane Bioreactors Used in Wastewater Treatment. *J. Membrane Sci.*, **284** (1-2), 17.

Lesjean, B.; Rosenberger, S.; Schrotter, J.-C. (2004) Membrane-Aided Biological Wastewater Treatment—An Overview of Applied Systems. *Membrane Technol.*, **8**, 5.

Li, N. N.; Fane, A. G.; Ho, W. S. W.; Matsuura., T. (2008) *Advanced Membrane Technology and Applications;* Wiley & Sons: Hoboken, New Jersey.

Meng, F.; Chae, S. R.; Drews, A.; Kraume, M.; Shin, H. S.; Yang. F. (2009) Recent Advances in Membrane Bioreactors (MBRs): Membrane Fouling and Membrane Materials. *Water Res.*, **43**, 1489.

Pan, J. R.; Su, Y. C.; Huang, C.; Lee, H. C. (2010) Effect of Sludge Characteristics on Membrane Fouling in Membrane Bioreactors. *J. Membrane Sci.*, **349**, 287.

Water Environment Federation; American Society of Civil Engineers; Environmental and Water Resources Institute (2009) *Design of Municipal Wastewater Treatment Plants,* 5th ed.; WEF Manual of Practice No. 8; ASCE Manual and Report on Engineering Practice No. 76; McGraw-Hill: New York.

Wilf, M.; Bartels, C.; Bloxom, D.; Christopher, J.; Festger, A.; Khoo, K.; Frenkel, V.; Hudkins, J.; Muller, J.; Pearce, G.; Reardon, R.; Royce, A. (2010) *The Guidebook to Membrane Technology for Wastewater Reclamation*; Balaban Publishers: Rehovot, Israel.

6.0 SUGGESTED READINGS

Gould, B.; Snodgrass, M.; Devine, J. (2005) Ultrafiltration of Municipal Wastewater Using an Immersed, Backflushable, Spiral Wound Membrane. *Proceedings of the 78th Annual Water Environment Federation Technical Exhibition and Conference* [CD-ROM]; Washington, D.C., Oct 29–Nov 2; Water Environment Federation: Alexandria, Virginia.

Mallevialle, J.; Odendaal, P.; Wiesner, M. (1995) *Water Treatment Membrane Processes;* McGraw-Hill: New York.

Stephenson, T. (2006) *Membrane Bioreactors for Wastewater Treatment;* IWA Publishing: London, U.K.

Zeman, L. J.; Zydney, A. L. (1996) *Microfiltration and Ultrafiltration: Principles and Applications;* Marcel Dekker: New York.

Chapter 3

Biological Process Fundamentals

(continued)

1.0 INTRODUCTION

This chapter expands on process fundamentals briefly discussed in Chapter 2 and introduces biochemical operations that occur in membrane bioreactor (MBR) systems, rate kinetics that govern substrate use, and biomass growth in bioreactors. The MBR process configurations to achieve biochemical oxygen demand (BOD), nitrogen, and phosphorus removal are also introduced. Differences in the characteristics of activated sludge among MBR and conventional activated sludge (CAS) systems; effects on membrane fouling, sludge viscosity, oxygen transfer, and sludge thickening; and removal of trace contaminants are also discussed.

2.0 PROCESS FUNDAMENTALS

The bioreactor component of an MBR is designed to biologically remove contaminants of concern, such as organic material and nutrients (i.e., nitrogen and phosphorus), from wastewaters. Although CAS and MBR systems differ in a number of aspects, overall biological process fundamentals for both systems are relatively similar. A brief review of biological process fundamentals is presented in the following section. A more detailed description of biological process fundamentals is presented in literature by Grady et al. (1999).

2.1 Biochemical Operations in Membrane Bioreactor Systems

Removal of contaminants of concern from wastewaters is predominantly achieved by a mixed population of bacteria that use organic material and nutrients as substrates for energy and cellular material production. Although complex and diverse, the mixed populations of microorganisms present in bioreactors can be grouped in a small number of categories depending on the types of substrates that they consume for energy and cellular material production and the environment required for their growth.

Organic material that can be biologically assimilated and/or oxidized is typically referred to as *biodegradable organic material*. Removal of biodegradable organic material is predominantly achieved by heterotrophic microorganisms under aerobic conditions. These microorganisms consume soluble organic material and oxygen for energy production (i.e., oxidation) and soluble organic material for cellular growth (i.e., assimilation). (It is important to note that nitrogen and phosphorus are also used as building blocks for cellular growth by all microorganisms.) In the process, carbon dioxide and more aerobic heterotrophic microorganisms are generated (see Figure 3.1a).

The removal of nitrogen is a two-step process (Figure 3.1b). The first step is achieved by autotrophic microorganisms under aerobic conditions. These microorganisms, which are typically referred to as *nitrifiers*, use ammonia and oxygen for energy production (i.e., oxidation) and use carbon dioxide for cellular growth (i.e., assimilation). In the process, nitrates, nitrites, and more nitrifiers are generated. Most of the nitrogen in the raw wastewater is typically present as ammonia, with organic nitrogen, nitrates, and nitrites present at lower concentrations. Particulate and soluble organic nitrogen are also present in wastewater. The second step is achieved by heterotrophic microorganisms in the absence of oxygen. These microorganisms, which are typically referred to as *ordinary denitrifiers*, use nitrates and nitrites as electron acceptor and organic material for energy and cellular growth. In the process, nitrogen leaves the system as nitrogen gas, and more denitrifiers are generated.

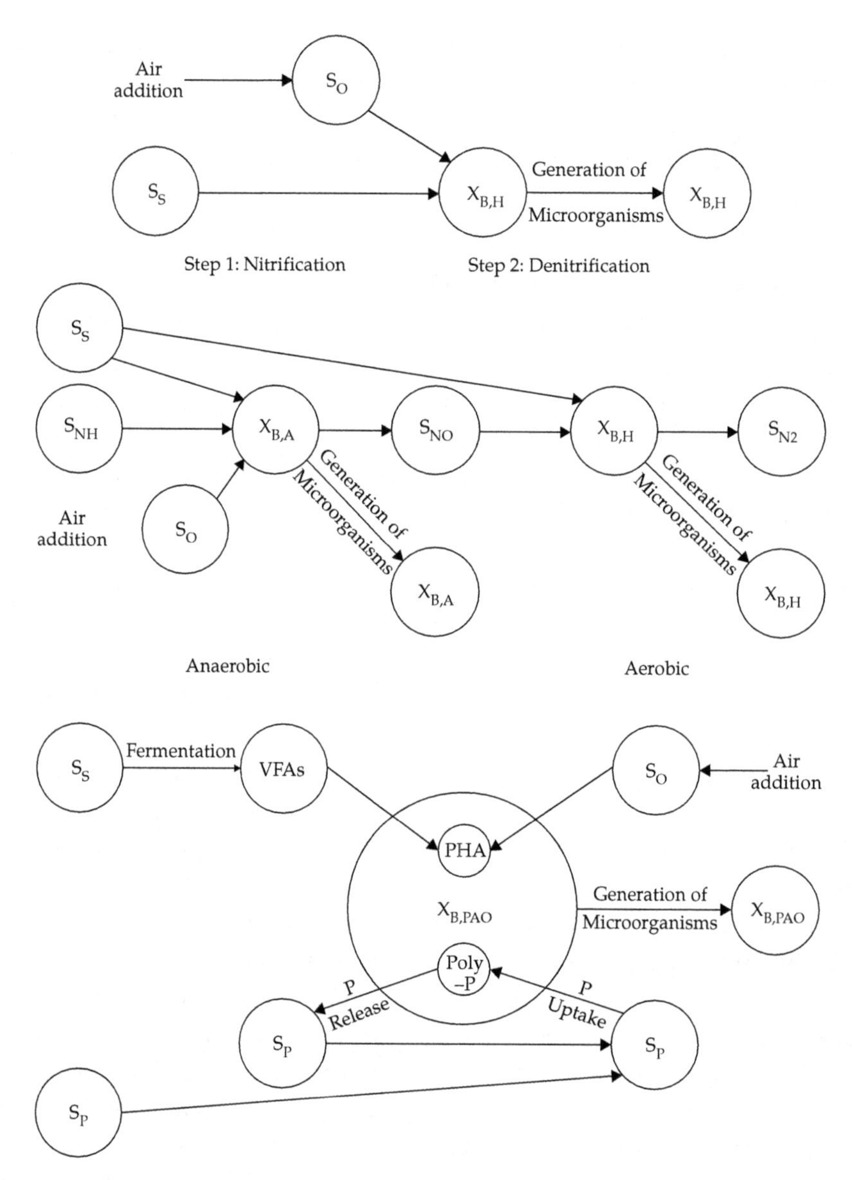

FIGURE 3.1 Schematic of contaminant removal by different groups of microorganisms (S_S = soluble organic material, X_S = particulate organic material, S_O = the soluble oxygen, $X_{B,H}$ = heterotrophic microorganisms, S_{NH} = ammonia, $X_{B,A}$ = autotrophic microorganisms, S_{NO} = nitrate and nitrite, S_{N2} = nitrogen, S_P = orthophosphate, $X_{B,PAO}$ = phosphorus accumulating microorganisms, PHA = poly-hydroxyl-alkanoates, Poly-P = polyphosphates, and VFAs = volatile fatty acids; all parameters expressed as concentrations (mass/volume) [adapted from Olsson and Newell (1999)]).

The removal of phosphorus is a relatively complex process (Figure 3.1c) involving the release and subsequent uptake of orthophosphate by phosphorus accumulating microorganisms when subjected to alternating anaerobic and aerobic conditions (Grady et al., 1999). These alternating conditions favor the growth of phosphorus accumulating organisms (PAOs) over other types of heterotrophic microorganisms. Approximately one-half of the phosphorus in wastewater is typically present as orthophosphate, with condensed phosphates (e.g., sugar phosphates and phospholipids) and organic phosphorus present at lower concentrations. During the anaerobic step, phosphorus accumulating microorganisms consume short-chain volatile fatty acids (VFAs) (i.e., acetate, propionate) and store poly-hydroxyl-alkanoates (PHAs), while soluble phosphates are released because of the breakdown of polyphosphate molecules. In the anaerobic stage, no growth of PAOs occurs. During the subsequent aerobic step, the stored PHAs are oxidized by PAOs for energy, which is then used for several cell functions including growth, maintenance, and replenishment of polyphosphate pool. In the aerobic zone, the released phosphate plus the phosphate initially present in wastewater is incorporated to the phosphorus accumulating microorganisms as polyphosphates. In the process, soluble organic material and oxygen are consumed for energy production, soluble organic material is consumed for cellular growth, and phosphorus accumulating microorganisms, which have a high phosphorus content, are generated. As discussed in Section 2.2, a portion of the microorganisms in a bioreactor is removed from the system (i.e., wasted), resulting in a removal of the stored phosphorus (Grady et al., 1999). It is important to note that the ability to effectively convert soluble organic material to VFAs is essential to the process. Fortunately, because of fermentation that occurs in sewers, some of the soluble organic material in wastewaters is in the form of fatty acids. However, some wastewaters lack sufficient fatty acids to promote phosphorus accumulation. For this reason, chemical phosphorus removal is often used to achieve high levels of biological phosphorus removal in MBRs (Metcalf and Eddy, 2003).

Some organic material and nutrients in wastewaters are present as particulate material and must first be hydrolyzed (i.e., solubilized) before they can be consumed. Soluble organic nitrogen must also be converted to ammonia (termed *ammonification*), and condensed phosphates and organic phosphorus to orthophosphate, before they can be consumed. These processes are mainly achieved by extracellular enzymes secreted by the microorganisms. Organic material and nutrients that cannot be biologically oxidized and/or assimilated are typically referred to as *inert* or *refractory*.

Over time, the mass of microorganisms generated, hereafter referred to as *biomass* (measured as volatile suspended solids [VSS]), decreases as some of the cellular material is converted to energy for cell maintenance while some of the microorganisms die or are consumed by higher microorganisms (i.e., predation). During these processes, which are typically grouped together and referred to as *endogenous decay*, particulate and soluble cellular material consisting of organic material and nutrients are released (i.e., lysis). Some of the released cellular material can be hydrolyzed to soluble material and reconsumed as described previously. It is important to note that microbial growth is also accompanied by the production of soluble microbial products. Combined material from microbial decay and cellular products account for most of the organic material present in the effluent from a wastewater treatment system.

2.2 Governing Rate Kinetics

Although mechanisms of organic material, nitrogen, and phosphorus consumption differ, growth of all the different groups of microorganisms responsible for the removal of these contaminants is governed by similar kinetic parameters.

Theoretically, the growth of microorganisms could proceed unimpeded as long as all of the substrates needed for growth were available. However, in closed systems such as the bioreactor in wastewater treatment processes, the concentration of some substrates may be low enough, or may become low enough as they are consumed, to limit the growth of microorganisms. For substrate-limiting conditions, the specific growth rate for microorganisms follows a Monod-type relationship as presented in Figure 3.2 and the following equation (Monod, 1949):

$$\mu' = \mu \frac{S}{(k_S + S)} \tag{3.1}$$

where μ' = specific growth rate (time^{-1}),

μ = maximum specific growth rate (i.e., for nonsubstrate-limited conditions) (time^{-1}),

S = concentration of substrate (mass/volume), and

k_S = half-saturation concentration (mass/volume).

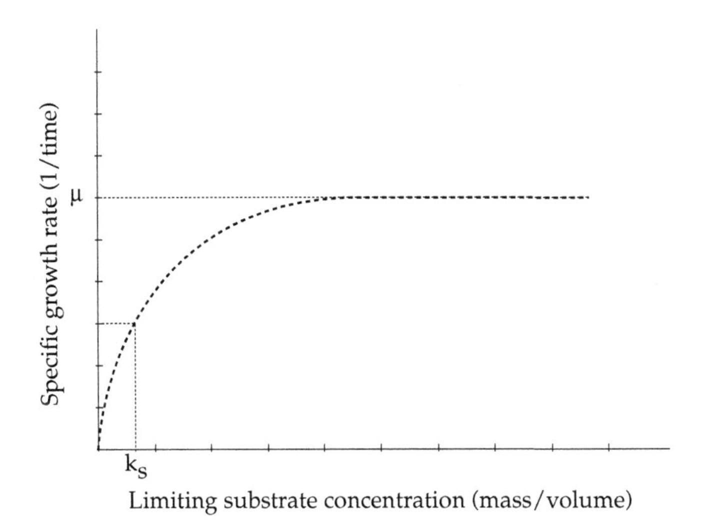

FIGURE 3.2 Effect of limiting substrate concentration on the specific growth rate.

It is important to note that the maximum specific growth rate and the half-saturation concentration are specific to a given group of microorganisms and for a given substrate.

Equation 3.1 assumes that microbial growth is limited by only one substrate. When more than one substrate is limiting, the specific growth rate can be estimated using the following equation:

$$\mu' = \mu \prod_l \left[\frac{S_l}{\left(k_{S_l} + S_l\right)} \right] \tag{3.2}$$

where the subscript l corresponds to the different limiting substrates.

From the specific growth rate, the rate of biomass growth can be estimated (first term on the left-hand side of eq 3.3). However, to estimate the net rate of biomass production, the rate of endogenous biomass decay (second term on the left-hand side of eq 3.3) must be subtracted from the rate of biomass growth as follows (Grady et al., 1999):

$$r_G = \frac{dX_B}{dt} = \mu \frac{S}{(k_S + S)} X_B - bX_B \tag{3.3}$$

where r_G = the net rate of biomass production (mass as VSS/volume·time),
X_B = the biomass concentration (mass as VSS/volume),
t = time (time), and
b = the endogenous decay rate (time^{-1}).

It is important to note that the biomass concentration and the endogenous decay rate are specific to a given group of microorganisms.

As discussed previously, substrate is consumed during biomass growth. The rate of substrate consumption can be estimated based on the rate of biomass growth and the yield coefficient, which is the ratio of mass of microorganisms formed to the mass of substrate consumed, as follows (Grady et al., 1999):

$$r_S = \frac{dS}{dt} = -\frac{1}{Y}\mu\frac{S}{(k_S + S)}X_B \tag{3.4}$$

where r_S = rate of substrate consumption (mass/volume·time), and
Y = yield coefficient (ratio of mass of microorganisms formed, measured as VSS, to the mass of substrate consumed).

It is important to note that the growth yield is specific to a given group of microorganisms and a given substrate. The rate of substrate consumption is negative because the substrate is consumed. Some of the particulate substrate released during endogenous decay can be hydrolyzed, contributing to the concentration of substrate in the system.

Although the membrane itself is not expected to affect the microbial community in an MBR, the greater retention of biomass, the higher mixed liquor suspended solids (MLSS) and the higher solids retention time (SRT). These are typical of an MBR compared to a CAS system and can result in differences in the composition of the microbial community between MBRs and CAS systems (Manser et al., 2005). Although a number of studies have reported differences in the composition of microbial communities and in the rate kinetics for different groups of microorganisms between MBRs and CAS systems (Cicek et al., 1999), others have reported limited to no differences between these systems (Gao et al., 2004; Manser et al., 2006). For this reason, rate kinetics that have been established for CAS are typically used to design MBRs. Typical values for rate kinetics of different groups of microorganisms in CAS systems can be found in literature by IWA Task Group on Mathematical Modelling for Design and Operation of Biological Wastewater Treatment (2000).

The effect of temperature on oxidation and assimilation of different substrates and the diffusion of substrates to the microorganisms is relatively complex. However, in general, the effect of temperature on different constants can be estimated using the following equation (Grady et al., 1999):

$$c_T = c_{20}\theta^{(20-T)} \tag{3.5}$$

where c = rate constant or coefficient,
θ = Arrhenius temperature correction factor, and
T and 20 = the temperature of interest and standard temperature of 20 °C, respectively.

It is important to note that the temperature correction coefficient is specific to a given group of microorganisms and a given substrate (Grady et al., 1999; Metcalf and Eddy, 2003). The rates in eqs 3.3 and 3.4 are typically reported for a temperature of 20 °C.

2.3 Biomass Growth and Substrate Consumption in Bioreactors

As presented in Sections 2.1 and 2.2, biomass growth, decay, and substrate consumption for the different groups of microorganisms are related and, therefore, must be considered simultaneously. A number of models, such as the IWA activated sludge models, have been developed to comprehensively simulate the biomass production and substrate consumption (IWA Task Group on Mathematical Modelling, 2000). A detailed description of these models is beyond the scope of the present discussion and is presented elsewhere (Grady et al., 1999; WEF et al., 2009). Nonetheless, development of the main bioreactor design parameters can be estimated by developing mass balances for biomass production and substrate consumption. Design parameters for a simple bioreactor configuration are presented in Figure 3.3. More complex configurations are introduced in Section 3.0.

For the simple bioreactor configuration presented in Figure 3.3, the mass balances for substrate consumption and biomass production are presented in eqs 3.6 through 3.9.

The solid–liquid separation step retains and returns the biomass to the bioreactor, allowing high concentrations of biomass to be maintained in the system. The return of biomass is typically referred to as *return activated sludge* (RAS). It should be noted

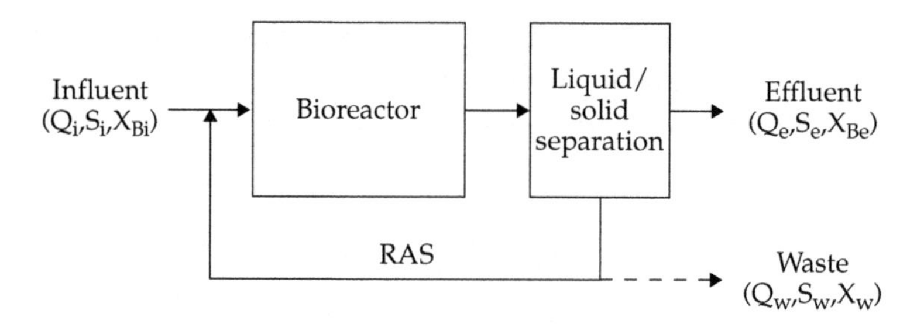

FIGURE 3.3 Schematic of a simple bioreactor configuration [Q = flow (volume/time); S = soluble substrate concentration (mass/volume); X_B = biomass concentration (mass/volume); X = solids concentration (mass/volume); and subscripts i, e, and w correspond to influent, effluent, and waste streams, respectively].

that the waste stream was omitted from mass balances presented in eqs 3.6 to 3.9 and can be calculated using eq 3.11. Eqs 3.6 through 3.9 are as follows:

$$\begin{matrix}\text{Change in mass} \\ \text{of substrate} \\ \text{in system}\end{matrix} = \begin{matrix}\text{Mass of substrate} \\ \text{entering system}\end{matrix} - \begin{matrix}\text{Mass of substrate} \\ \text{leaving system}\end{matrix} - \begin{matrix}\text{Rate of} \\ \text{substrate} \\ \text{consumption}\end{matrix} \tag{3.6}$$

$$\frac{dS}{dt} V_r = QS_O - QS - V_r r_S \tag{3.7}$$

$$\begin{matrix}\text{Change in} \\ \text{biomass} \\ \text{in system}\end{matrix} = \begin{matrix}\text{Biomass} \\ \text{entering system}\end{matrix} - \begin{matrix}\text{Biomass} \\ \text{leaving system}\end{matrix} - \begin{matrix}\text{Rate of} \\ \text{biomass} \\ \text{production}\end{matrix} \tag{3.8}$$

$$\frac{dX}{dt} V_r = QX_O - QX + V_r r_G \tag{3.9}$$

where V = volume of the bioreactor (volume),
Q = flow (volume/time), and
r and o = bioreactor and influent, respectively.

For steady-state conditions (i.e., $dS/dt = 0$), the mass balance for substrate consumption can be simplified and used to estimate the mean hydraulic retention time (HRT) required to achieve a desired amount of substrate consumption (i.e., S_i to S_e), based on a selected concentration of microorganisms (X_B) in the bioreactor (eq 3.10). Knowing the HRT (τ) and the flow to the system, it is possible to estimate the required bioreactor volume (V_r). It is important to note that the concentration of biomass in the influent (X_{Bi}) is typically considered to be negligible (i.e., all biomass is generated within the system). In addition, because the membrane used for liquid–solid separation in a membrane bioreactor can retain virtually all of the micoroorganisms in the system, the concentration of biomass in the effluent (X_{Be}) is also considered to be negligible. Eq 3.10 is as follows:

$$\tau = \frac{V_r}{Q} = \left(S_i - S_e\right)\frac{Y}{\mu}\frac{\left(k_S + S_e\right)}{S_e}\frac{1}{X_B} \tag{3.10}$$

where $\tau = HRT$ (T), and
i and e = influent and final effluent substrate concentrations, respectively.

It is important to note that eq 3.10 assumes that only one substrate is limiting.

Similarly, for steady-state conditions (i.e., $dX/dt = 0$), the mass balance for biomass can be simplified to estimate the mean microorganism retention time in the system, also commonly referred to as *SRT*, that is required to achieve a desired amount of substrate consumption (S_i to S_e) based on a selected concentration of microorganisms in the bioreactor (eq 3.11). Microorganisms produced in excess of the amount required to maintain the selected concentration of microorganisms in the bioreactor must be removed (i.e., wasted) from the system. Knowing the SRT (τ_c), the volume of the bioreactor (V_r), and the concentrations of microorganisms in the bioreactor (X_B) and waste stream (X_W) (note that the concentration of microorganisms in the waste stream depends on the location from where it is drawn [see Section 3.0]), it is possible to estimate the required waste flow (Q_W). The waste stream is typically referred to as *waste activated sludge*. It is important to note that the mass of residual substrate in the waste stream (S_W) is relatively small compared to the mass of residual substrate in the effluent (S_e) and is typically considered to be negligible. Eq 3.11 is as follows:

$$\tau_C = \frac{V_r X}{Q_W X_W} = \frac{1}{\mu[S_e/(k+S_e)] - b} = \frac{1}{\{(Y/X_B)[(S_i - S_e)/\tau] - b\}} \tag{3.11}$$

where τ_C = the SRT (time), and
r, w, and e = bioreactor, waste, and effluent, respectively, and
$(S_i - S_e)/\tau$ = an estimate of r_S.

The amount of waste sludge produced is also typically expressed in terms of the observed yield, which is the ratio of the biomass produced to the mass of substrate consumed (eq 3.12), and in terms of net mass of waste sludge, which is the product of the observed yield and the mass of substrate consumed (eq 3.13). The processing and disposal of excess microorganisms, typically referred to as *waste sludge*, accounts for a significant component of the overall cost associated with wastewater treatment. Therefore, it is desirable to minimize the production of waste sludge. This can be achieved, in part, by operating the bioreactor with a relatively high SRT, which favors endogenous decay (i.e., cell death and lysis, conversion of cellular material to energy for maintenance, and predation) (Drews and Kraume, 2005). Eqs 3.12 and 3.13 are as follows:

$$Y_{OBS} = \frac{Q_W X_W}{Q(S_i - S_e)} = \frac{Y}{1 + b\tau_C} \tag{3.12}$$

where Y_{OBS} = the observed yield (mass as VSS/mass of substrate consumed).

$$P_{X,VSS} = Y_{OBS} Q(S_i - S_e) \tag{3.13}$$

where $P_{X,VSS}$ is the wastage rate (mass as VSS/time).

As discussed in Section 2.1, oxygen is consumed by some groups of microorganisms. Unless replenished, the concentration of oxygen in the system can be depleted, limiting growth of these microorganisms. The amount of oxygen required can be estimated by developing an oxygen mass balance for the bioreactor. At steady state, the amount of oxygen required for the bioreactor is equal to the sum of the mass of oxygen consumed by the microorganisms for energy production (i.e., oxidation) and the mass of oxygen leaving the bioreactor as biomass (i.e., assimilation) (eq 3.14). For the removal of organic material, the mass of oxygen consumed by microorganisms is equivalent to the carbonaceous biochemical oxygen demand (CBOD) removed in the bioreactor. For nitrification, the mass of oxygen consumed is approximately 4.57 times the mass of ammonia (measured as nitrogen) removed in the bioreactor by nitrifiers for energy (Grady et al., 1999). Although no oxygen is consumed during denitrification, organic material is consumed and, therefore, overall oxygen demand for the removal of organic material is reduced. The mass of oxygen credit for denitrification is approximately 2.86 times the mass of nitrate (measured as nitrogen) removed in the bioreactor (Metcalf and Eddy, 2003). The mass of oxygen leaving the bioreactor as biomass is approximately 1.42 times the net mass of waste sludge (measured as VSS) (Metcalf and Eddy, 2003), as follows:

$$R_O = Q\left(S_{S_i} - S_{S_e}\right) + 4.57Q\left(S_{NH_i} - S_{NH_e}\right) - 2.86Q\left(S_{NO_i} - S_{NO_e}\right) - 1.42P_{X,VSS} \quad (3.14)$$

where R_O = the total oxygen required (M/T), and
S, NH, and NO = organic material, ammonia, and nitrate, respectively.

As discussed previously, the concentration of biomass in a bioreactor is measured as VSS. However, not all VSS in the bioreactor is biomass. Some VSS in the bioreactor originate from the influent particulate organic material, whereas the remainder originate from cell lysis. The fraction of active biomass in the mixed liquor can be estimated using eq 3.15, which is the ratio of the rate of biomass production to the total rate of solids accumulation in the system. Eq 3.15 is as follows:

$$F_X = \frac{r_G}{r_G + f_d bX + Q\left(X_{I_o}/V\right)} \quad (3.15)$$

where F_X = fraction of active biomass (–),
f_d = fraction of biomass that remains nonbiodegradable following decay (–), and
X_{I_o} = the concentration of nonbiodegradable VSS in the influent (mass as VSS/volume).

Application of the concepts introduced in Section 2.3 to MBR process design is discussed in Section 5.0 of Chapter 4.

3.0 PROCESS CONFIGURATIONS

An MBR is a combination of suspended-growth activated sludge biological treatment and membrane filtration equipment performing the critical solids–liquid separation function that is typically accomplished using secondary clarifiers (WEF, 2006). The key component of an MBR system is the low-pressure membrane system (i.e., microfiltration or ultrafiltration) that is used for solids–liquid separation. Because an MBR represents proprietary technology, there are differences in configuring the

MBR system and controlling membrane fouling. In some instances, design approach may vary among equipment suppliers. Although many leading membrane suppliers ignore the impact of biochemical transformations occurring in membrane tanks on process performance, assuming that membrane tanks are primarily for solids–liquid separation process, some transformations do occur in the membrane tank and should be taken into account in the design of an MBR system. Membrane bioreactor design approaches are discussed in Chapters 4 and 5.

An MBR is essentially a biological treatment process that uses the activated sludge process to remove soluble and particulate matter from the wastewater being treated. As in any activated sludge process, the key to proper operation is to create an environment that supports substrate use and biomass growth and to provide effective solids–liquid separation. Membrane bioreactors produce superior quality effluent that is essentially free of solids, colloidal materials, and bacteria. For enhanced nutrient removal, membrane bioreactors can be designed to achieve less than 3 mg/L total nitrogen and less than 0.05 mg/L total phosphorus (chemical addition) within one unit process, thereby eliminating the need for additional tertiary filters, denitrification filters, and tertiary phosphorus removal systems that are required downstream of CAS processes (Sen et al., 2008). For reuse applications requiring reverse osmosis use, membranes used in MBR systems can provide adequate pretreatment for reverse osmosis facilities, thereby eliminating the need for additional pretreatment facilities as required in CAS systems. Table 3.1 summarizes a typical MBR effluent quality for domestic wastewater applications. It should be noted that the majority of the values (i.e., total suspended solids [TSS], turbidity) presented in Table 3.1 reflect the effluent water quality specifications that are commonly used by membrane suppliers and design engineers. Although better effluent water quality can be achieved with MBRs, conservative values are used to account for declined water quality as a result of membrane fiber breakage and other integrity problems.

Elimination of secondary clarification in MBR systems allows operation of MBRs at much higher MLSS concentrations than CAS systems, which substantially reduce activated sludge basin volumes in an MBR system for a given biomass inventory or SRT. For a given process volume, an MBR process can operate at a longer SRT than a CAS plant, thereby reducing sludge (observed yield) production and wastage (WEF, 2006). Table 3.2 summarizes typical differences between CAS and MBR systems.

Biological process configurations of MBR and CAS systems are similar. Membrane bioreactor systems can be configured in numerous ways to promote the growth of

TABLE 3.1 Typical MBR effluent quality.

Parameter	Typical values
Five-day carbonaceous biochemical oxygen demand, mg/L	<5
TSS, mg/L	<1
Turbidity, nephelometric turbidity units	<0.2
Ammonia-nitrogen, mg/L	<1
Total nitrogen (with preanoxic zone), mg/L	<10
Total nitrogen (with preanoxic and postanoxic zones), mg/L	<3
Total phosphorus (biological phosphorus removal only), mg/L	<0.5
Total phosphorus (biological plus chemical phosphorus removal), mg/L	<0.1 (lower concentrations can be achievable)

TABLE 3.2 Typical differences between CAS and MBR systems.

	CAS	MBR
Fine screening	Not required	Required (typically 1 to 3 mm, uni-directional)
Activated sludge basin volume	Larger than MBR for a given sludge inventory or SRT	Smaller than CAS for a given sludge inventory or SRT
Typical MLSS in activated sludge basins	1500 to 3500 mg/L	4000 to 10 000 mg/L (activated sludge basins) 8000 to 18 000 mg/L (membrane tanks)
Typical RAS flow (% of the influent flow)	50 to 100	300 to 500
Wastage location	Secondary clarifiers	MBR tank (or the last aerobic zone)

microorganisms that can consume contaminants of interest (see Section 2.1), with the aim of achieving specific effluent quality requirements.

3.1 Fully Aerobic Membrane Bioreactor Systems

Carbon oxidation (CBOD removal) in activated sludge systems is a relatively easy task compared to other biochemical transformation (i.e., nitrification, denitrification, phosphorus removal) as long as noninhibitory biogenic substrate and the right environmental conditions are provided to heterotrophic organisms in activated sludge systems. In municipal wastewater application, most fully aerobic MBR systems have been designed to achieve combined CBOD removal and nitrification because using MBR for carbon oxidation only is generally cost-prohibitive, realizing that there are many cost-effective process solutions for CBOD removal (i.e., CAS systems, attached growth systems [i.e., trickling filters, rotating biological contactors], and aerobic lagoon systems). In addition, an SRT value of 10 days or more, which has commonly been used by MBR designers and membrane suppliers to minimize excessive production of biofoulants, allows MBR sludge to nitrify. *Fully aerobic system* herein refers to combined CBOD removal and nitrification system.

In a fully aerobic activated sludge process, biological oxidation of ammonia to nitrate can be achieved in combined CBOD removal and nitrification (single-stage) systems or in separate nitrification (two-stage) systems. The degree of nitrification in a combined, single-stage process depends on system SRT and temperature, provided that a population of nitrifying bacteria can be maintained. The basic approach to design of a suspended-growth nitrification process is the same as that for carbon oxidation and begins with determining an appropriate design SRT. The relatively slow growth rate of ammonia-oxidizing bacteria causes nitrifying systems to be slow to recover following process upsets because of low dissolved oxygen, depressed pH, and toxic inhibition, or to large changes in influent oxidizable nitrogen concentrations (WEF et al., 2009). A factor of safety is applied to the minimum necessary SRT for increased performance reliability based on consideration of variations in nitrogen loading, process performance requirements, and environmental factors (WEF et al., 2009).

A fully aerobic CAS process can easily be adapted to MBR by replacing secondary clarifiers with membrane tanks. The process flow schematic of a single-stage fully aerobic MBR is presented in Figure 3.4.

3.2 Combined Biochemical Oxygen Demand and Biological Nitrogen Removal Systems

Biological nitrogen removal is a two-step process that requires nitrification in an aerobic environment followed by denitrification in an anoxic environment. One prerequisite for good nitrogen removal is achieving complete nitrification and properly returning high nitrate containing internal recycle stream (internal mixed liquor recycle [IMLR] stream) to the anoxic zone(s). In MBR systems, RAS also provides nitrate enriched flows. However, RAS contains relatively high dissolved oxygen concentrations and returning

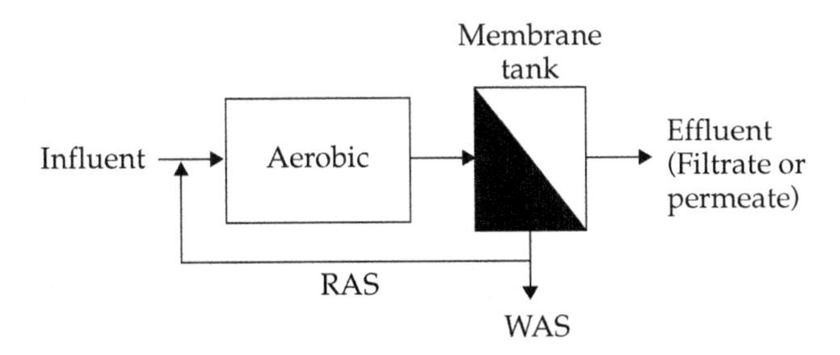

FIGURE 3.4 Process schematic of a fully aerobic MBR.

RAS to the anoxic zone reduces biological nitrogen removal performance of the system. A deoxygenation zone is typically designed to receive RAS flow prior to the anoxic zone to improve biological nitrogen removal. Approaches to improve nitrogen removal performance are further discussed in this chapter. The amount of refractory dissolved organic nitrogen (RDON) present in facility influent and created in biological treatment processes (WEF et al., 2009) limits the achievable effluent nitrogen concentration in activated sludge systems, including MBRs. Refractory dissolved organic nitrogen is organic nitrogen in soluble form that is not easily removed by biological treatment. Typical levels of RDON in wastewater treatment plant (WWTP) effluent range between 0.5 to 2 mg N/L, with most values in the range of 1 to 2 mg N/L (WEF et al., 2009). Therefore, it is not unusual for plants targeting low nitrogen levels (i.e., total nitrogen < 3 mg/L) to add supplemental carbon to the secondary treatment process. Because nitrifying organisms are slow-growing organisms compared to heterotrops, the design SRT should be properly selected to ensure nitrifying growth. In the combined nitrogen and nutrient removal systems, aerobic SRT is lower than total SRT because of inclusion of anoxic and anaerobic biomass. The aerobic SRT, defined as the mass of solids in the aerated portion of the reactors divided by mass of solid wastes per day, is, therefore, a key parameter for designing combined nitrogen and nutrient removal systems.

A variety of arrangements are available to achieve CBOD and biological nitrogen removal. Process selection and configuration are generally dictated by effluent nitrogen requirements, raw wastewater characteristics, and site-specific constraints. Combined BOD and nitrogen removal process configurations include

- Two-stage processes,
- Four-stage processes, and
- Step-fed configuration.

3.2.1 *Two-Stage Processes*

A two-stage system is a fairly simple and commonly used BOD and biological nitrogen removal configuration. Full-scale experiences with the two-stage process indicate that it can typically meet total nitrogen of 10 mg/L or less under favorable BOD/total Kjeldahl nitrogen (TKN) (i.e., 5:1). However, more stringent effluent total nitrogen (<5 or sometimes <3 mg/L) requires using more complex process configurations. The Modified Ludzack–Ettinger (MLE) process is a commonly used two-stage biological nitrogen removal process in CAS systems. An MLE process schematic is presented in Figure 3.5.

Incorporating IMLR increases both the denitrification rate and overall nitrogen removal efficiency and provides control over the fraction of nitrate removed through variation of the internal recycle ratio. Internal mixed liquor recycles are typically 100 to 200% of the influent flowrate. Two-stage biological nitrogen removal MBRs can be configured as numerous configurations. Some configurations may resemble MLE. In one configuration, an MBR tank simply replaces the secondary clarifier. In another other configuration, an MBR tank replaces the secondary clarifier and IMLR is eliminated knowing that RAS (300 to 500% of influent flow) can contain enough nitrate.

Membrane tanks have much higher dissolved oxygen concentrations (i.e., >6 mg/L) than do aerobic basins (i.e., 2 mg/L). Returning high dissolved oxygen containing RAS flow to the head of the anoxic zone can reduce denitrification and, hence, nitrogen removal performance of the system. To improve nitrogen removal performance in two-stage MBR systems, RAS is either returned to a deoxygenation zone or the aerobic zone and IMLR from the aerobic to the anoxic zone. The following

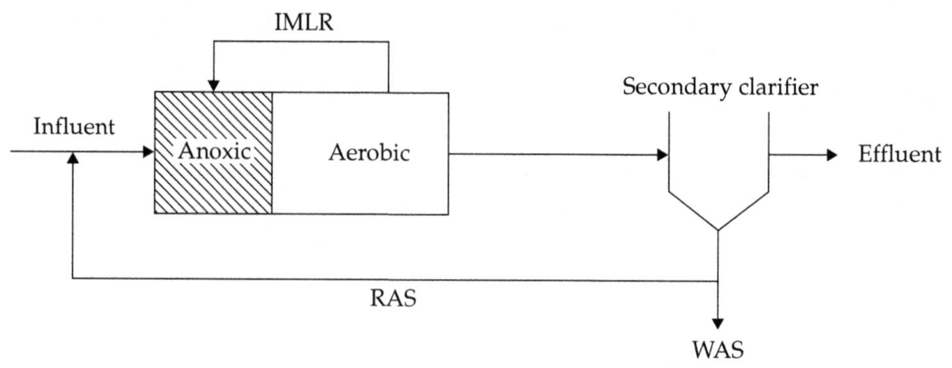

FIGURE 3.5 Modified Ludzack–Ettinger process schematic.

three configurations can be used in two-stage MBR systems (see Figures 3.6 through 3.9 for process schematics):

- *Configuration 1*—RAS flow to the aerobic zone and IMLR from aerobic to anoxic zone,
- *Configuration 2*—RAS flow to a deoxygenation zone and IMLR from aerobic to anoxic zone, and
- *Configuration 3*—RAS flow to a deoxygenation zone and no IMLR.

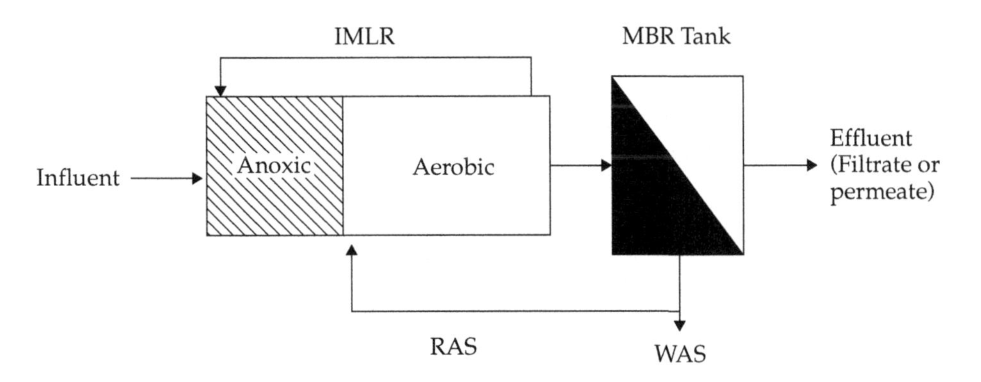

FIGURE 3.6 Two-stage MBR process—configuration 1.

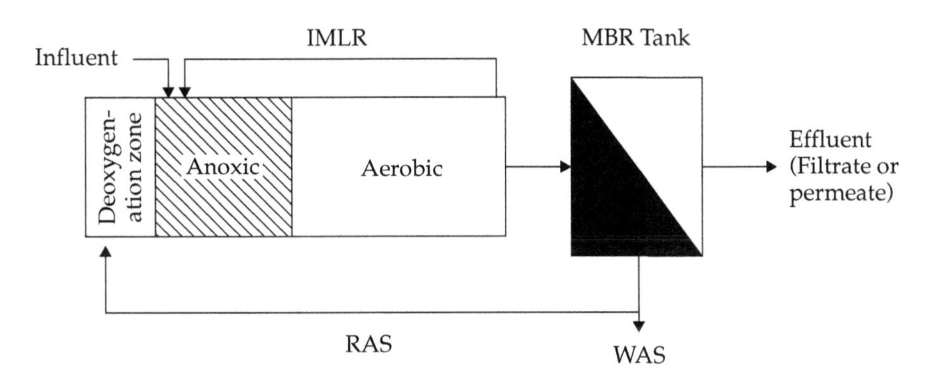

FIGURE 3.7 Two-stage MBR process—configuration 2.

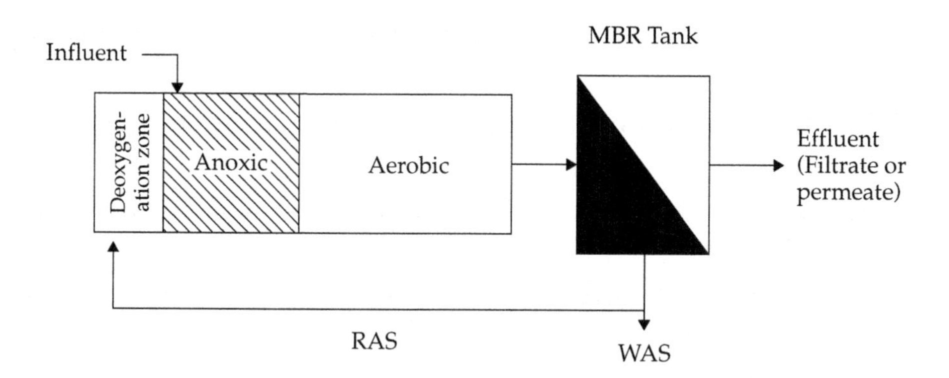

Figure 3.8 Two-stage MBR process—configuration 3.

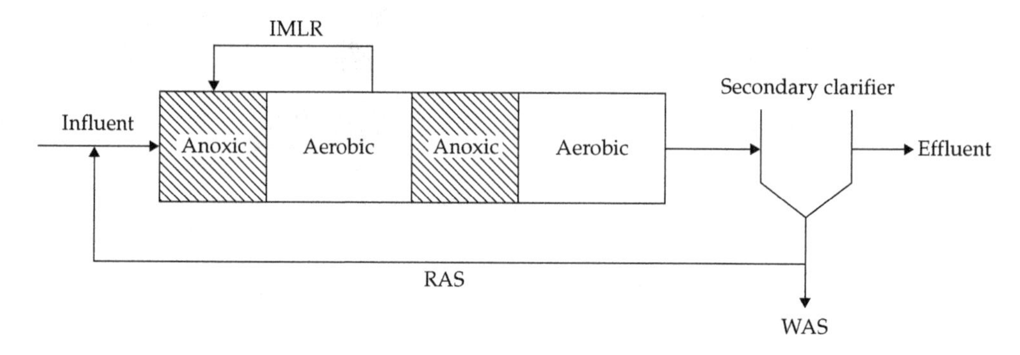

Figure 3.9 Process schematic of four-stage Bardenpho™.

3.2.2 *Four-Stage Processes*

The four-stage nitrogen removal process consists of a series of two anoxic and two aerobic zones, with recycling of mixed liquor from the first aerobic zone to the first anoxic zone at a rate as high as 4 to 6 times the influent flowrate (WEF et al., 2009). This process is intended to achieve more complete nitrogen removal than is possible with a two- or three-stage process. Complete denitrification cannot be attained with preaeration anoxic zones because part of the aerobic stage effluent is not recycled through the anoxic zone. The second anoxic zone provides for additional denitrification using nitrate produced in the aerobic stage as the electron acceptor and endogenous organic carbon or the supplemental organic carbon as the electron donor. The second (post-aeration) anoxic zone is capable of almost completely removing the nitrate in the aeration tank effluent, provided the size is adequate and supplemental carbon is added (WEF et al., 2009). The ability to successfully use the four-stage process to achieve an effluent concentration of total nitrogen as low as 2 to 4 mg/L depends on the ratio of oxidizable nitrogen to carbon in the influent to the activated sludge process and on the use of supplement carbon addition. Ekama et al. (1984) reported that influent COD/TKN must be no less than 13 to obtain complete denitrification (equivalent to BOD/TKN of approximately 6:1). A process schematic of four-stage Bardenpho™ used in CAS systems is illustrated in Figure 3.9.

This process can be modified for MBR systems. To maximize nitrogen removal performance, high dissolved oxygen containing RAS can be returned to the first aerobic zone with the use of IMLR, as depicted in Figure 3.10, or RAS is returned to a deoxygenation zone with or without IMLR, as presented in Figures 3.11 and 3.12.

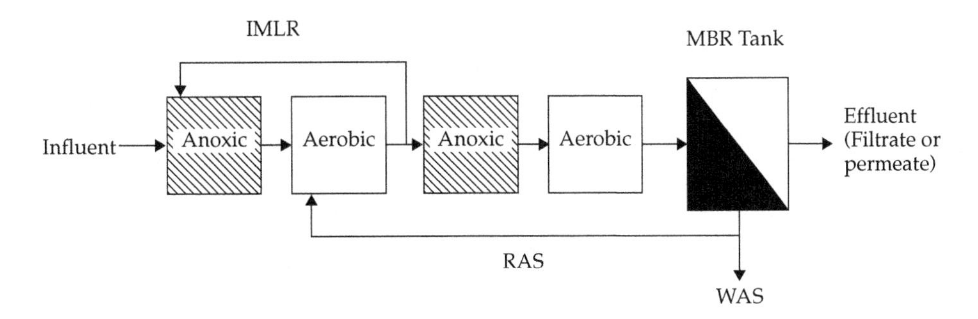

FIGURE 3.10 Process schematic of four-stage Bardenpho™—configuration 1.

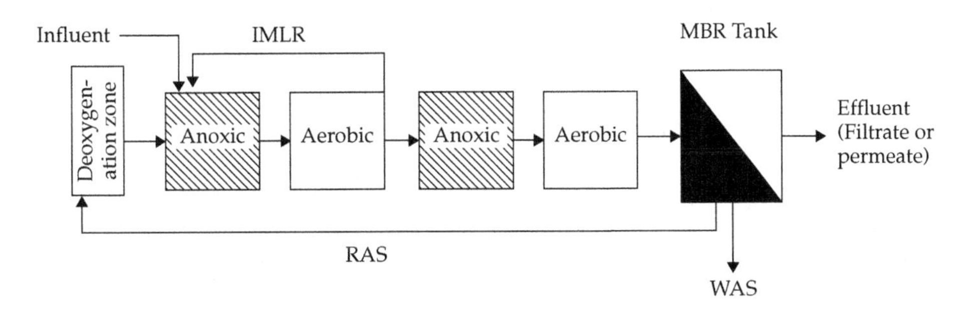

FIGURE 3.11 Process schematic of four-stage Bardenpho™—configuration 2.

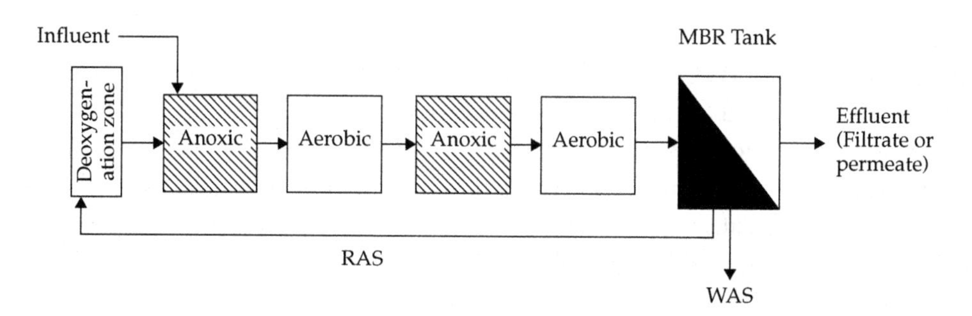

FIGURE 3.12 Process schematic of four-stage Bardenpho™—configuration 3.

3.2.3 *Step-Feed Membrane Bioreactor*

The step-feed membrane bioreactor process is essentially the same as conventional step-feed CAS, in which a portion of the influent to the process is fed to one or more points downstream of the head of the reactor to equalize the food-to-microorganism (F/M) ratio along the bioreactor tank and maximize nitrogen removal. The difference in step-feed nitrogen removal is that each of the feed points has an anoxic zone for nitrogen removal. As in conventional treatment, the primary benefits of the step-feed configuration include capacity improvement for a given volume or reduced reactor volumes for a specific capacity (WEF et al., 2009). In addition, for nitrogen removal, the step-feed process reduces or eliminates the need to recycle nitrate back to the anoxic zones. Nitrate is supplied to the anoxic zones from the upstream aerobic reactor directly, except for the first pass. In the first pass, the only nitrate supply is from the RAS stream. For this reason, nitrified recycle from the end of the first pass to the head of the first pass is sometimes provided to make best use of the available carbon (WEF et al., 2009). Step feed can also meet stringent effluent total nitrogen requirements (i.e., <5 mg/L) without providing supplemental BOD if the influent BOD/TKN is 5 or higher. Despite unique advantages of step-feed systems (i.e., reduced reactor volumes, reduced oxygen requirement), complexity of the flow split requires more skilled operation than other biological nitrogen removal (BNR) process configurations. Examples of step BNR MBR facilities include the Broad Run WWTP, Loudoun County, Virginia, which is designed with step-feed capability, and the Henderson WWTP, Henderson, Nevada (to be operated in step-feed mode). A process diagram of a typical step-feed MBR system with three feed points is shown in Figure 3.13.

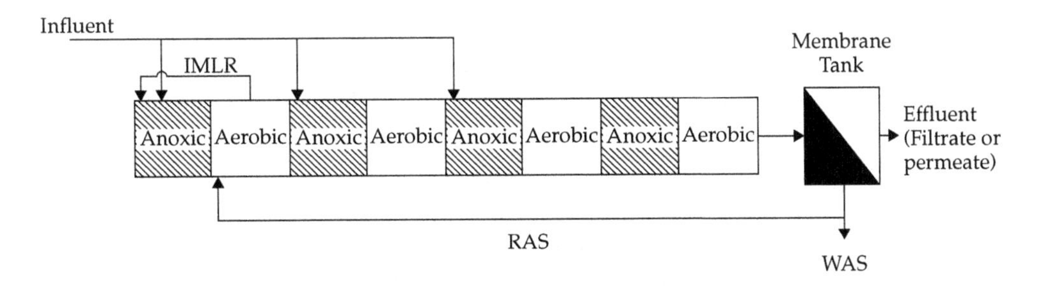

FIGURE 3.13 Process schematic of a step-feed MBR.

3.3 Combined Biological Nutrient (Nitrogen and Phosphorus) Removal Systems

The biological removal of both nitrogen and phosphorus in suspended-growth processes requires a process that has anaerobic (no molecular oxygen and oxidized form of nitrogen is present) zones, anoxic zones, and aerobic zones. The anaerobic zones are needed to give a competitive advantage to PAOs, anoxic zones for nitrogen removal, and aerobic zones for nitrification and phosphorus uptake. One key requirement to improve the nutrient removal performance is to restrict dissolved oxygen and nitrite and nitrate entries to the anaerobic zone(s) and minimize dissolved oxygen entry to the anoxic zone(s). Biological phosphorus removal is highly dependent on a lack of nitrate and nitrite in the anaerobic zone, so higher levels of denitrification in the nitrogen removal process benefit biological phosphorus removal (BPR) systems. (*Excess biological phosphorus removal* [EBPR] is a more commonly used term and often replaces *BPR*.) Conversely, a poorly performing nitrogen removal process can greatly decrease the performance of BPR (WEF et al., 2009). Sidestreams (i.e., centrate) from solids handling facilities of the biological nutrient system may increase both nitrogen and phosphorus loading to the activated sludge system. Process design should include effects of these streams. Because of the complexity of the system, BNR systems are being designed with whole-plant simulators based on IWA activated sludge models (IWA Task Group on Mathematical Modelling, 2000).

Combined biological nitrogen and phosphorus removal systems can achieve low levels of nutrients in the plant effluent (total nitrogen less than 5 mg/L; total phosphorus less than 1.0 mg/L). Because tighter effluent total phosphorus (i.e., <0.1 mg/L) cannot be reliably achieved through biological phosphorus removal only, a combination of biological, chemical, and physical methods are needed. When chemical addition is required for phosphorus removal, chemicals (metal salts; i.e., ferric, alum) can be directly injected upstream of membrane tanks. The microfiltration and ultrafiltration membranes used in MBRs are more efficient than conventional filtration (i.e., CAS system coupled with deep-bed media filtration) for removing phosphorus precipitates and MLSS. Membrane bioreactor systems with chemical addition can reliably meet effluent total phosphorus of less than 0.1 mg/L, whereas conventional filtration systems require careful operation and process optimization to meet stringent effluent total phosphorus requirements. As discussed previously, in suspended-growth secondary treatment systems, low levels of nitrogen in the effluent can be achieved as described earlier and are limited to the amount of RDON in the process effluent (typically between 1 and 2 mg N/L). It is not unusual for plants targeting low nitrogen levels to add supplemental carbon to the secondary treatment process.

A variety of arrangements are available to achieve nutrient removal. The activated sludge BNR configurations that are commonly used in BNR MBR systems are presented in the following sections. Complete lists of process configurations for BNR can be found elsewhere (Metcalf and Eddy, 2003; WEF et al., 2009).

3.3.1 Three-Stage Processes

Three-stage MBR BNR processes are a modified version of the three-stage CAS biological nutrient processes. Commonly used three-stage CAS biological nutrient processes include

- A2/O,
- University of Cape Town (UCT) and Modified UCT,
- Virginia Initiative Process, and
- The Johannesburg Process.

The description and process schematic of three-stage CAS BNR processes can be found in literature by Metcalf and Eddy (2003) and WEF et al. (2009). There are many possibilities for MBR three-stage biological nutrient processes. To improve BNR performance, three-stage MBR configurations should minimize the following: dissolved oxygen entry to anoxic and anaerobic zones and nitrate and nitrite entry to the anaerobic zone. This can be achieved in numerous ways. Example configurations of three-stage BNR MBR process are presented in Figures 3.14 through 3.16. These example configurations include the following:

- *Three-recycle configuration*—option 1 (RAS to deoxygenation zone, mixed liquor recycle [MLR] from aerobic zone to anoxic zone, anoxic MLR from anoxic to anaerobic zone);
- *Three-recycle configuration*—option 2 (RAS to aerobic zone, MLR from aerobic zone to anoxic zone, anoxic MLR from anoxic to anaerobic zone); and
- *Two-recycle configuration*—option 3 (RAS to deoxygenation zone, MLR from anoxic to anaerobic zone).

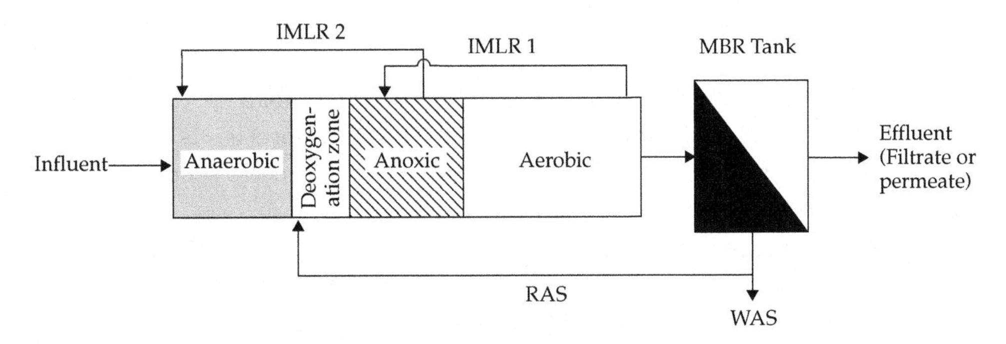

Figure 3.14 Three-stage MBR example—option 1.

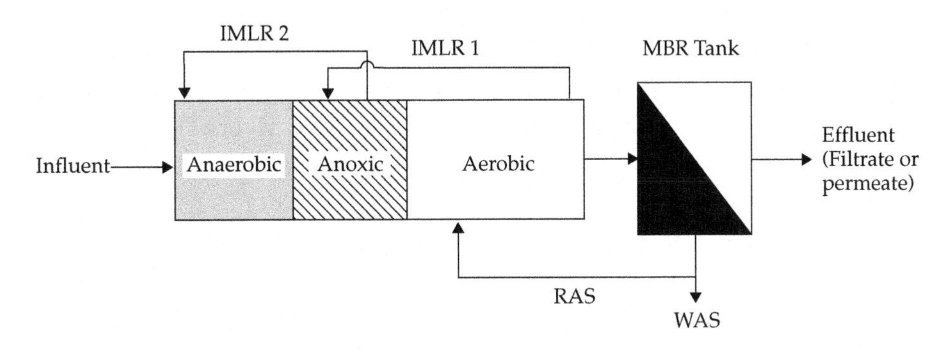

Figure 3.15 Three-stage MBR example—option 2.

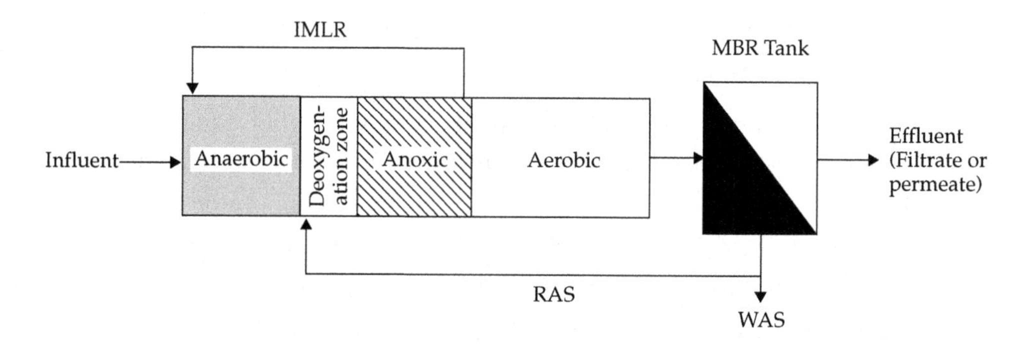

FIGURE 3.16 Three-stage MBR example—option 3.

3.3.2 *Five-Stage Processes*

Barnard (1976) proposed adding an anaerobic zone ahead of the four-stage Bardenpho™ process to allow combined BNR. The entire process is referred to as *five-stage Bardenpho™* or *modified Bardenpho™ process*. The process schematic of five-stage Bardenpho™ is illustrated in Figure 3.17.

In this configuration, influent wastewater combines with RAS in an anaerobic zone. The anaerobic zone gives a competitive advantage to PAOs where they store short-chain VFAs in the form of PHAs while releasing phosphate to the medium. In the subsequent anoxic zone, nitrate provided via internal recycle line is reduced to nitrogen gas. If the anoxic zone is sufficiently large, some PHA storage may take place subject to VFA availability and the exhaustion of nitrite and nitrate (Erdal et al., 2003). The flow then enters the aerobic zone where nitrification and phosphorus uptake take place. Internal mixed liquor recycle is typically between 200 and 500% of the average influent flow (WEF, 2005). In the second anoxic zone, further nitrate reduction is achieved. A small, final aerobic zone is designed to prevent anaerobic conditions for secondary phosphorus release. In Fort Myers Florida, five-stage Bardenpho™ CAS can meet effluent total nitrogen and total phosphorus of 3 and 0.5 mg/L, respectively.

Five-stage MBR BNR processes can be configured in numerous ways. The overall process schematic is similar to five-stage Bardenpho™ CAS. However, a consideration is given to minimize dissolved oxygen entry to the anoxic zone and dissolved oxygen and nitrate and nitrite to the anaerobic zone. The five-stage

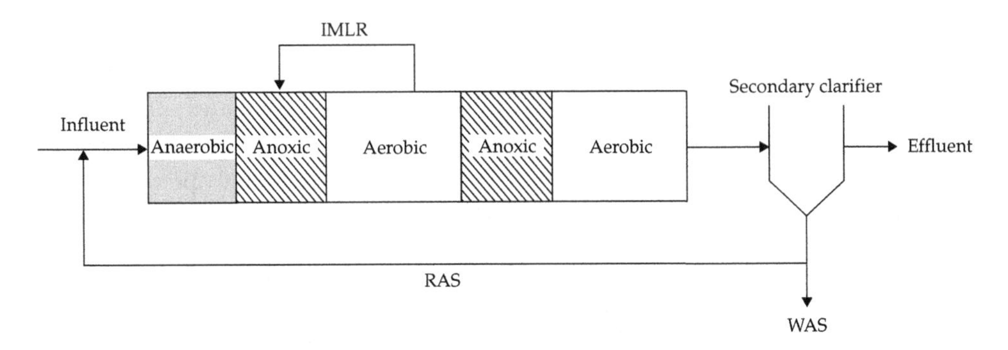

FIGURE 3.17 Process schematic of five-stage Bardenpho™.

process can be configured in several ways. Example configurations include the following:

- *Three-recycle configuration*—option 1 (RAS to deoxygenation zone, MLR from aerobic zone to the first anoxic zone, anoxic MLR from first anoxic to the anaerobic zone);
- *Three-recycle configuration*—option 2 (RAS to aerobic zone, MLR from aerobic zone to the first anoxic zone, anoxic MLR from first anoxic to the anaerobic zone); and
- *Two-recycle configuration*—option 3 (RAS to deoxygenation zone, MLR from first anoxic to the anaerobic zone).

Unlike the five-stage Bardenpho™ CAS, the five-stage MBR configuration does not require a small reaeration zone because reaeration can be achieved in the membrane tank.

4.0 CHARACTERISTICS OF ACTIVATED SLUDGE IN MEMBRANE BIOREACTOR SYSTEMS

The kinetic parameters and biochemical operations defined for activated sludge systems are directly applicable to MBR systems. Despite those similarities between CAS and MBR systems, mixed liquor characteristics are often different in each system. Such differences may be caused by differences in shear forces and mixing intensity acting on floc particles, MLSS concentration, and SRT of the systems. The operating

SRT is the master variable that influences growth characteristics of microorganisms, biodegradation of extracellular polymeric substances (EPS), formation of soluble microbial products (SMPs), oxygen requirement, sludge yield, and effluent water quality. It also has a strong effect on the selection of dominant microorganisms that are causing sludge bulking and foaming. This section discusses key operating parameters that influence activated sludge characteristics and membrane fouling and highlights the effect of high MLSS concentration on oxygen transfer and sludge thickening and dewatering.

4.1 Membrane Fouling and High Mixed Liquor Suspended Solids Concentration

Although high MLSS concentration in MBR systems is beneficial in reducing bioreactor volume, it might adversely affect membrane fouling. Membrane fouling is the systematic accumulation of suspended solids, colloids, cell debris, and macromolecules on the membrane surface or inside the pores, causing a reduction in membrane permeability. Shimizu et al. (1996) developed empirical relationships between critical flux and MLSS-concentration immersed hollow-fiber and tubular membranes (see Appendix B for a definition of *critical flux*). Their study indicated that increased MLSS reduces the permeate flux. However, the developed relationships also indicated that MLSS has minimal effect on the critical flux for MLSS greater than 5000 mg/L, which represents the lower end of the MLSS operating range in full-scale applications. Yamamoto et al. (1989) observed that MLSS concentration affected filtration resistance at concentrations exceeding 30 000 mg/L, which is much higher than that maintained in a typical MBR operation. Similarly, Sato and Ishii (1991) observed a nearly linear correlation between filtration resistance and the concentration of MLSS in the range of 20 000 to 29 000 mg MLSS/L. Trussell et al. (2001) described an upper MLSS limit, ranging from 24 000 to 34 000 mg/L, for which a sharp viscosity increase resulted in severe membrane fouling. Current practice indicates that immersed MBR systems have typically operated with MLSS concentrations between 8000 and 12 000 mg/L in the membrane tanks, with occasional operation between 15 000 and 18 000 mg/L (Crawford et al., 2000). Mixed liquor suspended solids concentration appears to have little effect on filtration resistance in MLSS concentration ranges of 5000 and 20 000 mg/L. This indicates that membrane fouling is not linked to MLSS concentration in a typical MBR operating range; rather, it is dependent on activated sludge characteristics (i.e., EPS, SMP content, viscosity) and operating conditions. These practical limits have been set to ensure adequate air transfer in the bioreactor and are discussed later in this section.

4.2 Effects of Operating Solids Retention Time, Extracellular Polymeric Substances, and Soluble Microbial Products on Membrane Fouling

As stated previously, SRT dictates MLSS concentration and physiological state of microorganisms in activated sludge basins, effluent quality, sludge wastage and oxygen requirements, and EPS and SMP contents, which can greatly influence membrane fouling (Ahmed et al., 2007; Han et al., 2005; Jinsong et al., 2006). *Extracellular polymeric substances* and *SMP* both refer to a variety of macromolecules such as polysaccharides, nucleic acids, proteins, and other polymeric compounds that are produced by secretion, shedding of cell surface material, and cell lysis. Definitions of *EPS* and *SMP* are vague in the literature, and some researchers use the two terms interchangeably. One commonly used definition of *EPS* is that it is a solid matrix that is bound or floc-associated, whereas SMPs represent soluble macromolecules (Meng et al., 2009). Extracellular polymeric substances play a key role in the formation of activated sludge flocs (Menniti et al., 2009) and also provide stability to the flocs in high shear environments (Mikkelsen et al., 2002). Soluble microbial products are released in the solution during substrate use and biomass decay. In MBRs, these soluble compounds interact with the membrane surface and can clog or adsorb to membrane pores (Menniti et al., 2009). High concentrations of EPS has been correlated with increased fouling and cake resistance (Meng et al., 2009). Recent studies have shown that SMPs are also significant foulants in MBRs (Meng et al., 2009; Pan et al., 2010).

Studies in the literature agree that low SRTs (2 days) have been shown to dramatically increase fouling rates because EPS is secreted by the microorganisms. Similarly, Trussell et al. (2006) showed that membrane fouling and SMP productions were well correlated in the investigated SRT of 2 to 10 days, where the fouling rate was 10 times lower at 10-day SRT than 2-day SRT. Drews et al. (2008) concluded that as SRT increases to 20 to 30 days, the relevance of SMP for filtration resistance and fouling decreases. A recent study indicated EPS, or SMP (the two terms are used interchangeably here), is the significant foulant and that its content varied in two MBRs that are operated under identical SRT, MLSS, and operating flux (Rosenberger et al., 2006). However, despite numerous studies, actual mechanisms leading to SMP formation regarding SMP properties are not clearly understood and findings are often contradictory. This can be partly attributed to the differences in plant configurations, operating conditions (i.e., SRT, MLSS), and membrane operating conditions (flux, backwash cleaning interval) (Drews et al., 2008).

In early MBR applications, SRTs of the system were selected to be long enough (>40 days) to minimize EPS excretions and provide better treatment for the slowly biodegradable material (WEF, 2006). Currently, it is recommended that MBR process designers select a suitable SRT to ensure complete nitrification for a given temperature and to maintain manageable MLSS concentration for proper oxygen transfer rather than minimizing EPS and/or SMP production. Fouling caused by these microbial products and other foulants can be controlled by automated in situ membrane cleaning (WEF, 2006).

4.3 Effects of Mixed Liquor Suspended Solids Concentration and Viscosity on Oxygen Transfer

The relatively high MLSS and small reactors associated with MBRs require that more oxygen be transferred per unit reactor volume. However, oxygen-transfer kinetics may decrease with increased MLSS concentration and viscosity. High viscosity can reduce the interfacial area for oxygen transfer by increasing the rate of bubble coalescence (Schwarz et al., 2006). In fact, several MBR studies reported that oxygen demand can exceed the volumetric capacity of the aeration system at high MLSS concentrations (>13 000 mg/L) (Schwarz et al., 2006).

The influence of mixed liquor constituents on aeration capacity can be quantified by the alpha (α) factor (see Appendix B for a definition of α). Because physical features and operating conditions of aeration equipment can vary, the relationship between MLSS and the α value is system-specific (Schwarz et al., 2006). The effect of MLSS concentration on oxygen transfer rates and, particularly, on the α value has been addressed in the three studies (Figure 3.18).

Figure 3.18 indicates that the aeration α value decreases exponentially with increased MLSS concentration while the rate of decrease is system-specific. For example, using curve fit developed by Krampe and Krauth (2003) in Figure 3.18 suggests that increasing activated sludge MLSS concentration from 3 to 6 g/L will reduce the aeration α from 0.77 to 0.59, resulting in 23% reduction in α. Similarly, if an α value of no less than 0.5 is targeted, the activated sludge basin MLSS concentration should not exceed 8 g/L per Krampe and Krauth (2003). Current design practice considers MLSS to be closer to 8000 to 10 000 mg/L to ensure reasonable oxygen transfer efficiency (WEF et al., 2009).

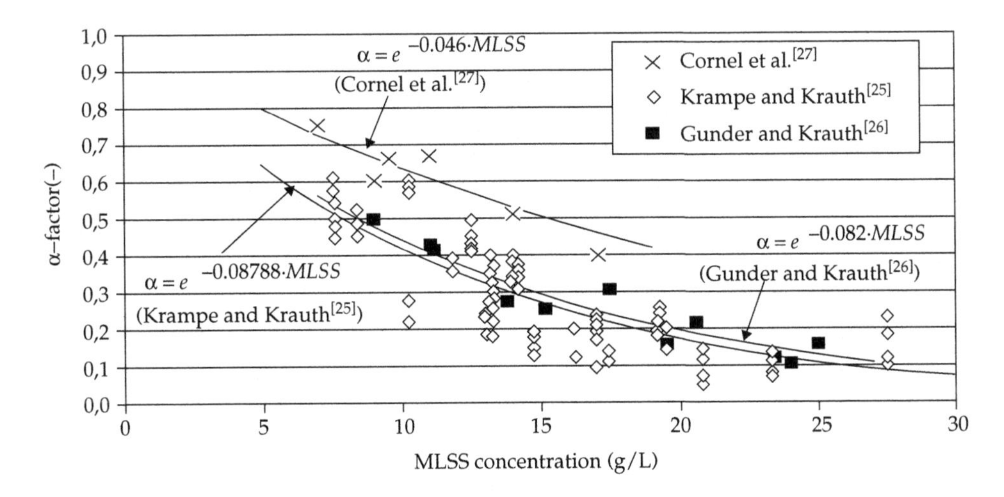

Figure 3.18 Mixed liquor suspended solids—aeration α factor relationship in fine-bubble diffuse aeration systems (adapted from WERF, 2002).

4.4 Sludge Thickening Characteristics of Membrane Bioreactor Sludge

Conventional activated sludge systems select for microorganisms that are well flocculated, which should correlate to good thickening. Conversely, MBRs retain all microorganisms and small flocs regardless of their settling properties (Schwarz et al., 2006). Therefore, MBRs generate smaller and weaker flocs. In addition, MBR flocs are subjected to erosion because of a higher MLSS concentration and increased shear (Schwarz et al., 2006). Yi et al. (2006) showed that measured median particle diameter was much smaller in MBR systems (10 μM) than CAS systems (120 μM) operated under identical conditions. Smaller flocs observed in MBR systems increase the exposed surface area. As a result, MBR sludge may thicken poorly or require high doses of polymer.

4.5 Sludge Bulking and Foaming

An MBR represents an activated sludge process and, therefore, is subject to similar sludge bulking and foaming problems occurring in activated sludge systems. By replacing secondary clarification with membrane separation, MBR systems reduce filamentous sludge bulking and settling issues. In activated sludge, substrate can be used quickly by microorganisms in certain instances, which may lead to low dissolved oxygen conditions. Low dissolved oxygen bulking is caused by filamentous bacteria such as *Sphaerotilus natans*, Type 021N and Type 1701. Expanding the capacity of the aeration system is a solution to overcoming low dissolved oxygen bulking (Rittmann and McCarty, 2001).

Membrane bioreactor systems are generally operated with longer SRTs than CAS, thereby making MBRs more susceptible to low F/M bulking. Low F/M bulking is typically caused by *Microthrix parvicella*, Type 0041, Type 0092, and Type 0581, and *Haliscomenbacter hydrosis* (Jenkins et al., 2003). Because SRT of the system cannot be reduced in an attempt to eliminate low F/M bulking, the best approach is to design a selector reactor in front of the aerobic reactor. Membrane bioreactor systems used in BNR and EBPR systems have nonaerated zones that could act like a selector and favor floc-forming bacteria, thereby reducing low F/M bulking (Jenkins et al., 2003).

Another common problem in activated sludge systems is the formation of foam and scum on the surface of aeration basins. *Nocardia* sp. (most common), *Microthrix parvicella* (less common), and Type 1863 (rarest) can cause activated sludge foaming (WEF, 2002). *Nocardia* growth is typically associated with fats, oils, and grease present in treated water; warmer water temperatures; and longer SRTs. Longer SRTs typically used in MBR systems exacerbate foaming problems. Providing sufficient freeboard to accommodate foam accumulation and providing surface wasting facilities to remove foam from the surface are proactive approaches to deal with foam. Foam management strategies are discussed further in Chapter 5.

4.6 Observed Sludge Yields in Conventional Activated Sludge and Membrane Bioreactor Systems

Process configurations and operating conditions can significantly affect the observed sludge yield. Therefore, it is crucial to consider the process configuration and operating conditions when comparing the observed sludge yield for CAS and MBR systems.

4.6.1 *Observed Sludge Yields under Similar Process Configuration and Operating Conditions*

For MBR and CAS systems having primary clarifiers and operated under identical conditions (i.e., SRT temperature, feed composition, and process configuration), similar observed sludge yields are expected. Activated sludge model simulations (BioWin [EnviroSim Associates Ltd., Hamilton, Ontario, Canada]; GPS-X [Hydromantis Environmental Software Solutions, Inc., Hamilton, Ontario]; etc.) from the IWA Task

Group on Mathematical Modelling (2000) also indicate similar observed sludge yields for MBR and CAS systems under identical operating conditions. Research studies indicate that the unique floc structure of MBRs may promote favorable conditions for the growth of predator organisms (protozoa and metazoa), which may affect sludge yield. However, predation by protozoa and metazoa and their contributions to the overall biomass decay in activated sludge systems were not clearly understood (Van Loosdrecht and Henze, 1999). In addition, current process models are not capable of predicting their effect on process performance and sludge yield. Therefore, current process design practice typically ignores such factors.

4.6.2 *Observed Sludge Yields and Solids Production under Dissimilar Process Configuration and/or Operating Conditions*

Membrane bioreactor systems are typically operated with longer SRTs than CAS systems, resulting in lower sludge yield in MBR systems than CAS systems because of increased endogenous respiration (biomass decay) at higher SRTs. Despite increasing oxygen requirements, operating activated sludge plants at longer SRTs is advantageous because it reduces sludge yield and, thus, operating costs associated with solids disposal cost.

Membrane bioreactor systems are often designed without primary clarifiers. This reduces the overall solids production and, hence, the observed sludge yield because more organic material is oxidized to carbon dioxide in MBR systems than CAS systems.

4.7 Removal of Trace Contaminants in Conventional Activated Sludge and Membrane Bioreactor Systems

Membrane bioreactor systems completely retain the biomass, which allows proper control of SRT in MBR systems. Longer SRT operation, more diverse population (Van Loosdrecht and Henze, 1999) and excellent MLSS removal in MBR systems produce better effluent quality for gross surrogate contaminants (i.e., TSS, BOD). Because WWTPs are not designed to remove specific trace contaminants (i.e., pharmaceuticals, personal care products, endocrine disrupting compounds), the fate of these compounds in activated sludge systems is not widely understood. The literature studies agreed that biodegradation/biotransformation and adsorption are the two main mechanisms for removal of trace contaminants in activated sludge systems (Drewes et al., 2006; Holbrook et al., 2002). Trace organics are typically highly soluble and relatively small compounds (i.e., <1000 daltons). Moreover, they can freely pass the membranes (molecular weight cutoff of ultrafiltration membranes used in MBRs is typically 3000 to 400 000 daltons) used in MBR systems, thereby indicating that those membranes have no direct effect on the removal of trace contaminants (Snyder et al., 2007). Trace contaminants also have a low Henry's constant, which eliminates removal of these compounds via volatilization (i.e., air stripping).

The extent of removal for many trace contaminants was found to be similar for MBR and CAS systems, suggesting that biodegradation is the principal removal mechanism (Clara et al., 2005; Drewes et al., 2006). Conversely, Takigami et al. (2000) and Holbrook et al. (2002) showed that the primary removal mechanism for naturally occurring hormones is adsorption rather than biodegradation.

Because SRT and temperature are the master parameters that dictate biodegradation and removal of contaminants, CAS and MBR removal capabilities should be compared using similar SRTs and temperatures to eliminate bias. Erdal et al. (2009) compiled data from the literature studies to compare CAS and MBR in terms of trace contaminant removal efficiencies under similar operating conditions. The literature concurred that MBR and CAS systems operated under identical SRT and temperature perform similarly for the removal of the majority of trace contaminants (Drewes et al., 2006; Stephenson and Oppenheiemer, 2007). Conversely, it was found that certain trace contaminants with a high octanol-water partition coefficient (log $Kow > 3$) (i.e., 17a-estradiol, 17b-estradiol, fluoxetine, gemfibrozil) (Holbrook et al., 2002; Mansell et al., 2005) can be more effectively removed in MBR than CAS systems. As discussed previously, MBR has smaller floc sizes and particle diameters than CAS systems. This enhances adsorption of hydrophobic compounds (reflected in high *Kow*) on MBR MLSS. Because membranes used in MBR systems are effective barriers to solids particles, trace contaminants adsorbed by MLSS can be effectively removed in MBR systems. This explains why MBR exhibits better removal efficiencies for hydrophobic compounds (i.e., fluoxetine, gemfibrozil, 17b-estradiol) than CAS.

5.0 REFERENCES

Ahmed, Z.; Cho, J.; Lim, B.-R.; Song, K.-G.; Ahn, K.-H. (2007) Effects of Sludge Retention Time on Membrane Fouling and Microbial Community Structure in a Membrane Bioreactor. *J. Membrane Sci.*, **287** (2), 211.

Cicek, N.; Franco, J.; Suidan, M. T.; Urbain, V.; Manem J. (1999) Characterization and Comparison of a Membrane Bioreactor and Conventional Activated

Sludge System in Treatment of Wastewater Containing High Molecular Weight Compounds. *Water Environ. Res.*, **71**, 64.

Crawford, G.; Thompson, D.; Lozier, J.; Daigger, G.; Fleischer, E. (2000) Membrane Bioreactors—A Designer's Perspective. *Proceedings of the 73rd Annual Water Environment Federation Technical Exposition and Conference* [CD-ROM]; Anaheim, California, Oct 14–18; Water Environment Federation: Alexandria, Virginia.

Clara, M.; Kreuzinger, N.; Strenn, B.; Gans, O; Kroiss, H. (2005) The Solids Retention Time—A Suitable Design Parameter to Evaluate the Capacity of Wastewater Treatment Plants to Remove Micropollutants. *Water Resour. Res.*, **39** (1), 97.

Cornel, P.; Wagner, M.; Krause, S. (2003) Investigation of Oxygen Transfer Rates in Fullscale Membrane Bioreactors. *Water Sci. Technol.*, **47** (11), 313–319.

Drewes, J. E.; Hemming, J.; Schauer, J. J.; Sonzogni, W. (2006) *Removal of Endocrine Disrupting Compounds in Water Reclamation Processes;* Water Environment Research Foundation: Alexandria, Virginia.

Drews, A.; Kraume, M. (2005) Process Improvement by Application of Membrane Bioreactors. *Chem. Eng. Res. Des.*, **83**, 276.

Drews, A.; Vocks, M.; Bracklow, U.; Iversen, V.; Kraume, M. (2008) Does Fouling in MBR Depend on SMP? *Desalination*, **231**, 141.

Ekama, G. A.; Marais, G. V. R. (1984) Biological Nutrient Removal. In *Theory, Design and Operation of Nutrient Removal Activated Sludge Processes;* Water Research Commission: Pretoria, South Africa.

Erdal, U. G.; Erdal, Z. K.; Randall, C. W. (2003) The Competition between PAOs (Phosphorus Accumulating Organisms) and GAOs (Glycogen Accumulating Organisms) in EBPR (Enhanced Biological Phosphorus Removal) Systems at Different Temperatures and the Effects on System Performance. *Water Sci. Technol.*, **47** (11), 1.

Erdal, U. G.; Shyamasundar, V.; Schimmoller, L.; Daigger, G. T. (2009) Linear and Non-Linear Models to Predict Removal Efficiencies of Compounds of Emerging Concern (CECs) During Wastewater Treatment. *Proceedings of the 82nd Annual Water Environment Federation Technical Exhibition and Conference* [CD-ROM]; Orlando, Florida, Oct 10–14; Water Environment Federation: Alexandria, Virginia.

Gao, M. C.; Yang, M.; Li, H. Y.; Wang, Y. M.; Pan, F. (2004) Nitrification and Sludge Characteristics in a Submerged Membrane Bioreactor on Synthetic Inorganic Wastewater. *Desalination,* **170**, 177.

Grady, L. C. P.; Diagger, G. T.; Lim, H. C. (1999) *Biological Wastewater Treatment,* 2nd ed.; Marcel Dekker: New York.

Gunder, B.; Krauth, K. (1999) Replacement of Secondary Clarification by Membrane Separation—Results with Tubular, Plate, and Hollow Fiber Modules. *Water Sci. Technol.*, **40** (4–5) 311–320.

Holbrook, R. D.; Novak, J. T.; Grizzard, T. J.; Love, N. G. (2002) Estrogen Receptor Agonist Fate during Wastewater and Biosolids Treatment Process: A Mass Balance Analysis. *Environ. Sci. Technol.*, **36**, 4533.

IWA Task Group on Mathematical Modelling for Design and Operation of Biological Wastewater Treatment (2000) *Activated Sludge Models ASM1, ASM2, ASM2d, and ASM3;* IWA Publishing: London, U.K.

Jenkins, D.; Richard, M. G.; Daigger, G. T. (2003) *Manual on the Causes and Control of Activated Sludge Bulking and Foaming and Other Solids Separation Problems,* 3rd ed.; IWA Publishing: London, U.K.

Jinsong, Z.; Chuan, C. H.; Jiti, Z.; Fane, A. G. (2006) Effect of Sludge Retention Time on Membrane Bio-Fouling Intensity in a Submerged Membrane Bioreactor. *Sep. Sci. Technol.*, **41** (7), 1313.

Krampe, J.; Krauth, K. (2003) Oxygen Transfer into Activated Sludge with high MLSS Concentrations. *Water Sci. Technol.*, **47** (11), 297.

Mansell, B.; Peterson, J.; Tang, C.; Horvath, R. W.; Stahl, J. F. (2005) Membrane Bioreactor (MBR) Piloting at a Water Reclamation Plant in Los Angeles County. *Proceedings of the 78th Annual Water Environment Federation Technical Exhibition and Conference* [CD-ROM]; Washington, D.C., Oct 29–Nov 2; Water Environment Federation: Alexandria, Virginia.

Manser, R.; Gujer, W.; Siegrist, H. (2005) Membrane Bioreactor Versus Conventional Activated Sludge System: Population Dynamics of Nitrifiers. *Water Sci. Technol.*, **52**, 417.

Manser, R.; Gujer, W.; Siegrist, H. (2006) Decay Process of Nitrifying Bacteria in Biological Wastewater Treatment Systems. *Water Res.*, **40**, 2416.

Meng, F.; Chae, S. R.; Drews, A.; Kraume, M.; Shin, H. K.; Yang, F. (2009) Recent Advances in Membrane Bioreactors (MBRs): Membrane Fouling and Membrane Material. *Water Res.*, **43**, 1489.

Menniti, A.; Kang, S.; Elimelech, M.; Morgenroth, E. (2009) Influence of Shear on the Production of Extracellular Polymeric Substances in Membrane Bioreactors. *Water Res.*, **43** (17), 4305.

Metcalf and Eddy, Inc. (2003) *Wastewater Engineering: Treatment and Reuse*, 4th ed.; McGraw-Hill: New York.

Mikkelsen, L. H.; Mascarenhas, T.; Nielsen, P. H. (2002) Key Parameters in Sludge Dewatering: Testing for the Shear Sensitivity and EPS Content. *Water Sci. Technol.*, **46** (10), 105.

Monod, J. (1949) The Growth of Bacterial Cultures. *Annu. Rev. Microbiol.*, **3**, 371.

Olsson, G.; Newell, B. (1999) *Wastewater Treatment Systems: Modelling, Diagnostics and Control;* IWA Publishing: London, U.K.

Pan, J. R.; Su, Y. C.; Huang, C. P. (2010) Characteristics of Soluble Microbial Products in Membrane Bioreactor and its Effect on Membrane Fouling. *Desalination*, **250**, 778.

Rittmann, B.; McCarty, P. (2001) *Environmental Biotechnology: Principles and Applications*; McGraw-Hill: New York.

Rosenberger, S.; Laabs, C.; Lesjean, B.; Gnirss, R.; Amy, G.; Jekel, M.; Schrotter, J.-C. (2006) Impact of Colloidal and Soluble Organic Material on Membrane Performance in Membrane Bioreactors for Municipal Wastewater Treatment. *Water Res.*, **40**, 710.

Sato, T.; Ishii, Y. (1991) Effects of Activated Sludge Properties on Water Flux of Ultrafiltration Membrane used for Human Excrement Treatment. *Water Sci. Technol.*, **23** (7–9), 1601.

Schwarz, A. O.; Rittmann, B. E.; Crawford, G. V.; Klein, A. M.; Daigger, G. T. (2006) Critical Review on the Effects of Mixed Liquor Suspended Solids on Membrane Bioreactor Operation. *Sep. Sci. Technol.*, **41**, 1489.

Sen, D; Randall, C. W. (2008) Improved Computational Model (AQUIFAS) for Activated Sludge, IFAS and MBBR Systems, Part I: Semi-Empirical Model Development. *Water Environ. Res.*, **80** (5), 439.

Shimizu, Y.; Okuno, Y.; Uryu, K.; Ohtsubo, S.; Watanabe, A. (1996) Filtration Characteristics of Hollow Fiber Microfiltration Membranes used in Membrane Bioreactor for Domestic Wastewater Treatment. *Water Res.,* **30** (10), 2385.

Snyder, S. A.; Wert, E. C.; Lei, H.; Westerhoff, P.; Yoon, Y. (2007) *Removal of EDCs and Pharmaceuticals in Drinking and Reuse Treatment Processes.* American Waterworks Association Research Foundation: Denver, Colorado.

Stephenson, R.; Oppenheimer, J. (2007) *Fate of Pharmaceuticals and Personal Care Products through Municipal Wastewater Treatment Processes;* Water Environment Research Foundation: Alexandria, Virginia.

Takigami, H.; Taniguchi, N.; Matsuda, T.; Yamada, M. (2000) The Fate and Behavior of Human Estrogens in a Night Soil Treatment Process. *Water Sci. Technol.,* **42** (7–8), 45.

Trussell, R. S.; Merlo, R.; Adham, S.; Gagliardo, P.; Trussell, R. R. (2001) Understanding the Upper Limit for Suspended Solids in the Operation of Submerged Membrane Bioreactors (SMBR). *Proceedings of 74th Annual Water Environment Federation Technical Exhibition and Conference* [CD-ROM]; Atlanta, Georgia, Oct 13–17; Water Environment Federation: Alexandria, Virginia.

Trussell, R. S.; Merlo, R. P.; Hermanowicz, S. W.; Jenkins, D. (2006) The Effect of Organic Loading on Process Performance and Membrane Fouling in a Submerged Membrane Bioreactor Treating Municipal Wastewater. *Water Res.,* **40**, 2675.

Van Loosdrecht, M. C. M.; Henze, M. (1999) Maintenance, Endogeneous Respiration, Lysis, Decay and Predation. *Water Sci. Technol.,* **39** (1), 107.

Water Environment Federation (2002) *Activated Sludge,* 2nd ed.; WEF Manual of Practice No. 9; Water Environment Federation: Alexandria, Virginia.

Water Environment Federation (2005) *Biological Nutrient Removal (BNR) Operation in Wastewater Treatment Plants;* WEF Manual of Practice No. 29; McGraw-Hill: New York.

Water Environment Federation (2006) *Membrane Systems for Wastewater Treatment;* McGraw-Hill: New York.

Water Environment Federation; American Society of Civil Engineers; Environmental and Water Resources Institute (2009) *Design of Municipal Wastewater Treatment Plants,* 5th ed.; WEF Manual of Practice No. 8; ASCE Manual and Report on Engineering Practice No. 76; McGraw-Hill: New York.

Water Environment Research Foundation (2002) *Membrane Technology: Feasibility of Solid/Liquid Separation in Wastewater Treatment.* Water Environment Research Foundation: Alexandria, Virginia.

Yamamoto, K.; Hiasa, M.; Mahmood, T.; Matsuo, T. (1989) Direct Solid-Liquid Separation using Hollow Fiber Membrane in an Activated Sludge Aeration Tank. *Water Sci. Technol.,* **21** (4–5, part 1), 43.

Yi, T; Harper, W. F., Jr; Holbrook, D.; Love, N. G. (2006) Role of Particle Size and Ammonium Oxidation in Removal of 17-Ethinyl Estradiol in Bioreactors. *J. Environ. Eng.* (Reston, Virginia), Nov, 1527.

6.0 SUGGESTED READINGS

Barker, D. J.; Stuckey, D. C. (1999) A Review of Soluble Microbial Products (SMP) in Wastewater Treatment Systems. *Water Res.,* **33** (14), 3063.

Cicek, N.; Macomber, J.; Davel, J.; Suidan, M. T.; Audic, J.; Genstet, P. (2001) Effect of Solids Retention Time on the Performance and Biological Characteristics of a Membrane Bioreactor. *Water Sci. Technol.,* **43** (11), 43.

Chang, I. S.; Le-Clech, P.; Jefferson, B.; Judd, S. (2002) Membrane Fouling in Membrane Bioreactors for Wastewater Treatment. *J. Environ. Eng.* (Reston, Virginia), **128** (11), 1018.

Daigger, G. T.; Randall, C. W.; Waltrip, G. D.; Romm, E. D. (1987) Factors Affecting Biological Phosphorus Removal for the VIP Process, A High-Rate University of Cape Town Type Process. In *Biological Phosphate Removal from Wastewaters;* Ramadori, R., Ed.; Pergamon Press: Oxford, U.K.

Le-Clech, P.; Chen, V.; Fane, T. A. G. (2006) Fouling in Membrane Bioreactors used in Wastewater Treatment. *J. Membrane Sci.,* **284** (1–2), 17.

Madigan, M. T.; Martinko, J. M.; Parker, J. (2000) *Brock Biology of Microorganisms,* 9th ed.; Prentice Hall: New York.

Stephenson, T.; Judd, S.; Jefferson, B.; Brindle, K. (2000) *Membrane Bioreactors for Wastewater Treatment;* IWA Publishing: London, U.K.

Wagner, M.; Cornel, P.; Krause, S. (2002) Efficiency of Different Aeration Systems in Full-Scale Membrane Bioreactors. *Proceedings of the 75th Annual Water Environment Federation Technical Exposition and Conference* [CD-ROM]; Chicago, Illinois, Sept 28–Oct 2; Water Environment Federation: Alexandria, Virginia.

Chapter 4

Membrane Bioreactor Process Design

(continued)

1.0 INTRODUCTION

Building on fundamentals provided in previous chapters, this chapter presents methodology and considerations for completing membrane bioreactor (MBR) process design. This chapter also addresses design of all significant components in an MBR flow sheet, including pretreatment, biological treatment, membrane separation, post-treatment, and residuals treatment. Sections specifically devoted to design for energy optimization and process simulation are also included.

2.0 MEMBRANE BIOREACTOR PROCESS OVERVIEW

An MBR system typically consists of equipment for pretreatment, suspended growth biological treatment, membrane separation, post-treatment, and solids treatment or handling.

2.1 Pretreatment

Pretreatment is critical to the long-term performance of an MBR plant and to prolonging membrane life. Design and selection of the pretreatment system is application-specific and covers a wide range of processes that may include grit and grease removal, coarse and fine screening, primary clarification, and equalization or other methods for peak flow management. Design considerations of downstream processes for the various methods of pretreatment will be discussed in Section 4.0.

2.2 Biological Process

The choice to elect biological processes is driven mainly by wastewater characteristics and the desired effluent quality. All process configurations require aeration, recirculation, and mixing. Design considerations for the biological treatment process of an MBR system are addressed in Section 5.0. Chapter 3 provides further details on the fundamental aspects of biological wastewater treatment processes.

2.3 Membrane Separation Process

Although a number of different configurations of membrane filtration equipment are commercially available, the general principles of membrane operation are the same. Permeate extraction is achieved by inducing a pressure differential across the membrane, and the solids concentration is controlled by recirculation of sludge to deconcentrate the membrane surface. Membrane cleaning is controlled by inducing shear at the membrane surface, combined with periodic introduction of cleaning chemicals. Section 6.0 provides further details on design considerations for the membrane separation process of an MBR facility.

2.4 Post-Treatment

Membrane bioreactor post-treatment considerations are similar to conventional activated sludge (CAS) treatment systems, except that MBR effluent offers some advantages for post-treatment design. Section 7.0 provides additional details on MBR post-treatment.

2.5 Residuals Treatment

In general, solids treatment and handling for an MBR system is similar to that of a CAS system, with some additional design considerations. These are summarized in Section 8.0.

3.0 DEFINITION OF MEMBRANE BIOREACTOR DESIGN BASIS

3.1 Design Basis of Biological Process

The information required to establish a biological process design for an MBR is the same as that required for a CAS plant. It includes wastewater flowrates, wastewater characteristics, environmental conditions, and treatment objectives. Other design considerations, such as the existing site plan, space constraints, hydraulic profile, and so on, may also influence design and selection of the biological process.

3.2 Design Basis of Membrane Separation Process

The design basis of a membrane separation process should include the information needed to determine the membrane surface area and the size of associated equipment. The following information is typically provided in a design basis:

- Design flows for the membrane filtration equipment, including all unique flow conditions and their duration. Typical flow conditions include average daily flow, maximum month flow, maximum week flow, maximum day flow, and peak hourly flow. Details related to future phased expansion or initially reduced flow should also be considered.

- The range of mixed liquor temperature, including different temperatures for different flow conditions, if applicable. Minimum and maximum temperatures for all flow conditions are critical to membrane surface area selection.
- Mixed liquor suspended solids (MLSS) concentration in the membrane tank, including design ranges for different flow conditions, if applicable.

Additional information specific to the application that may affect the nature of the mixed liquor include fats, oils, and grease (FOG) concentrations, quality and location of solids processing sidestream return flows caused by potential polymer effects, type and dose of metal salt coagulants or other chemicals, and the presence of solvents or other constituents that may affect the membrane filtration performance.

3.3 Equipment Redundancy

Equipment redundancy is incorporated into the design of treatment plants to maintain reliability of continued operation and to reduce or eliminate downtime while continuously meeting water quality standards. The level of redundancy is typically dictated by regulatory requirements and may be exceeded by customer or designer preference. For some unit operations in an MBR system, the design engineer may include redundancy even though it may not be required by regulation; for example, redundancy of fine-screening equipment ensures protection of membrane assets at all times. Redundancy requirements should be included as part of the plant design basis because they can influence equipment design and selection. For an MBR system, *redundancy* is typically defined as the flow that must be treated with a certain number of membrane trains out of service for a prescribed duration, typically referred to as an $N–x$ design, where N is the total number of membrane trains in the plant and x is the number of membrane trains out of service. If the specified redundancy allows for one or more membrane trains to be offline indefinitely, this is commonly referred to as $N+x$, where N is the number of duty membrane trains and x is the number of standby membrane trains.

Redundancy requirements can be less stringent for sewer mining or scalping applications because of the opportunity to divert the flows to a downstream facility provided that the downstream facility can accommodate the additional load. For privately owned and operated facilities, redundancy may be critical to meeting contractual treatment requirements.

4.0 PRETREATMENT

4.1 Grit and Grease Removal

Grit and grease removal can be an essential part of the MBR process. Grit can be detrimental to the MBR system in two ways: (1) accumulation of grit within the aeration basin and/or the membrane tank, which must be removed, and (2) potential abrasion and/or damage to the membranes. Any grit removal system associated with the CAS process can be used with similar considerations (e.g., potential adverse effects on nutrient reduction caused by oxygen addition when using aerated grit upstream of biological systems using anaerobic or anoxic zones). Chopper pumps may also be considered to homogenize the flow before it reaches grit removal and fine-screen facilities, although communication with the screening vendor is required to verify compatibility.

Grease removal should be considered based on an understanding of influent conditions. Most domestic wastewater treatment systems do not include specific grease removal systems because the grease content of the flow is minor. Primary clarifiers would serve as a means to remove minor amounts of grease. However, if grease concentrations exceed typical values of 50 to 100 mg/L (Metcalf and Eddy, 2003) encountered in domestic wastewater, or if an atypical concentration of mineral oil is anticipated, grease removal should be provided to protect the downstream fine screens from clogging and the membranes from fouling. Comprehensive testing of FOG in the influent may be implemented to help understand the type and quantity of FOG and the potential interplay between the biological system and membranes (e.g., biodegradability, solubility). Detailed discussions should be held with the membrane manufacturer to assess the required level of pretreatment. In addition, using a periodic hot water wash can be effective in removing grease that has adhered to the surface of fine screens.

4.2 Screening

Effective screening is critical for MBR systems to protect membranes from damage and to maximize their service life. Experience has shown that properly designed, installed, operated, and maintained fine-screening equipment reduces the need for membrane maintenance and extends membrane life. Without proper screening, trash, hair, and fibrous materials can accumulate in the system and damage the membranes. Consequently, all MBRs require influent fine screens with 1- to 3-mm omnidirectional openings (i.e., opening size is uniform in all directions). Screens must have no potential to carryover solids (e.g., a single-pass design where the screen elements are not rotating back through the screened effluent with a potential for solids carryover). Chapter 5 provides further details on commonly applied fine-screen equipment types.

The fine screening required for an MBR increases the cost of screening equipment and the amount of screenings produced. According to data from various research and operational facilities, use of screen openings between 0.75 mm and 2 mm will result in a screenings concentration of 13 to 25 mg/L (based on raw wastewater); with screen openings smaller than 0.75 mm, the screenings concentration can be as high as 94 mg/L (based on raw wastewater) (Côté et al., 2006; Van der Roest et al., 2002). Screenings typically must be washed and compacted, which adds to both capital and operation and maintenance (O&M) costs for the system, although the additional capital cost is typically less than 3% of the total MBR facility cost (Côté et al., 2006). One of the reasons for washing and compacting, in addition to reducing the disposal costs of hauling the waste solids off-site, is to return the organic material to the mainstream process. If the screenings are washed for the purposes of disposal and the wash water is returned to the process, the majority of biochemical oxygen demand (BOD) will be returned to the system. This step can be critical for successful nutrient reduction if the plant influent has low BOD. Typical removal rates of total suspended solids (TSS) and BOD through fine screens are presented in Table 4.1.

To minimize loading to the fine screens and reduce the quantity of fine screenings produced, dual screening may be considered. An upstream screen with larger openings followed by traditional handling and disposal procedures can be used to remove large debris. The downstream fine screens will, therefore, receive

Table 4.1 Typical removal rates by fine screening (based on raw wastewater) (Metcalf and Eddy, 2003).*

Type of screen	Size of openings, mm	Percent TSS removed	Percent BOD removed
Fixed parabolic	1.6	5–30	5–20
Rotary drum	0.25	25–45	25–50

*The actual removal rate will depend on the nature of the wastewater collection system and the wastewater travel time.

wastewater with less solids content, which reduces the amount of screenings to be washed and compacted. Disposal of fine screenings to the solids processing facilities is an alternative method of handling that eliminates the need for compacting equipment and disposal containers. With this method, the screenings are passed through grinders to protect downstream solids processing equipment. One concern with this method of disposal is the loss of organics noted previously, which may adversely affect nutrient removal if the plant influent has low organic concentration.

Alternative locations for screening include post primary clarifiers (where applicable) between the aeration basins and the membrane tanks or sidestream screening of MLSS. Locating fine screens downstream of primary clarifiers results in a significant reduction of screening material; some research shows that the screenings concentration could be as little as 3% of the screenings concentration of raw wastewater depending on the screen-opening size (Van der Roest et al., 2002).

Additional screening of mixed liquor can also be used in tandem with influent screening. One option is to provide screening of mixed liquor by locating the screens between the aeration basins and membrane tanks. This option is best suited for retrofitting existing plants where the membrane tanks are located away from aeration basins and MLSS must be pumped to membrane tanks. This layout allows for the use of screens with a larger opening at the head of the plant. A disadvantage of this screening method is the need to screen the entire flow to the membranes, which could be up to 5 times the volume of the influent flow.

A less common method of screening the mixed liquor is sidestream screening, which is used in combination with influent screens to mitigate the risk of bypassing the influent screen or to remove debris that may blow into the basins. In this configuration, a separate screen is used to treat a sidestream of mixed liquor equal to 25 to 100% of the influent flow in combination with a 2- to 3-mm influent screen. The sidestream screening could be applied to either a portion of the flow from the MLSS recycle in denitrification systems or the return activated sludge (RAS) stream. Solids captured by sidestream screening are typically combined with influent screenings handling and disposal.

4.3 Primary Clarification

For large MBR plants, typically greater than 20 ML/d, the tendency is to consider the addition of primary clarifiers to reduce the load to the MBR, similar to the benefits of designing large CAS treatment plants. However, it is important to evaluate the effect of adding primary clarifiers completely. Traditionally, the principal reason for using a primary clarifier has been load reduction to (1) reduce the power use associated with aeration; and (2) reduce the biological tank volume. Because MBRs are operated at longer solids retention times (SRTs) (i.e., sufficient to achieve complete nitrification), primary clarifiers help maintain a higher volatile fraction of mixed liquor and reduce the accumulation of inert suspended solids. For an MBR, aeration requirements include process air and membrane scour air, with the volume of scour air being approximately equal to the process air volume. With a reduction in organic load, process aeration requirements would be reduced; however, the scour air requirements would not change. Consequently, the actual power reduction would be associated only with the process air and, although absolute energy savings would be the same, it would represent a smaller fraction of the overall aeration power than for a conventional plant. This concept is illustrated in the example provided in Table 4.2, which shows that while the use of primary clarifiers yields a 23% reduction in process air requirements, the total air requirements only reduce by 12% because of the contribution of membrane air scour.

Reducing the organic load to an MBR has other energy- and O&M-related benefits. Decreasing the organic and solids loadings on the MBR process means that, for a given flowrate, the MBR process can operate at a lower MLSS concentration. This, in turn, has two key benefits: (1) reduced membrane fouling, leading to longer intervals between cleanings and longer membrane life, and (2) greater oxygen transfer efficiency because of the lower organic loading rate and/or the increased alpha factor, resulting in lower power consumption by the process aeration blower.

TABLE 4.2 Aeration requirements for 68 ML/d domestic MBR with and without primary clarifiers (based on Black & Veatch biological model simulation).

	Process air, Nm^3/h (scfm*)	Membrane scour air, Nm^3/h (scfm)	Total air, Nm^3/h (scfm)
Without primary clarifier	40 100 (23 600)	38 200 (22 500)	78 300 (46 100)
With primary clarifier	30 900 (18 200)	38 200 (22 500)	69 100 (40 700)
Air reduction because of primary clarifier addition	9200 (5400)	0 (0)	9200 (5400)

*SCFM = standard cubic feet per minute (scfm × 1.699 = Nm^3/h defined at 20° C and 1 atm).

Use of primary clarifiers also adds treatment considerations with respect to the plant's overall energy balance. Inclusion of a primary clarifier results in a two-sludge system, which provides an opportunity to use anaerobic digestion. The energy associated with gas production and increased solids reduction from using anaerobic digestion may be beneficial in the long run and, therefore, may outweigh the costs associated with implementing a primary clarifier.

From a design standpoint, use of primary clarifiers affects other process elements. There is an opportunity to locate fine screens downstream of the primary clarifiers, which would significantly reduce screenings production and, therefore, screenings handling. Primary clarification is also effective at removing finer grit particles that may make it past fine screens and the grit removal process. In addition, primary clarifiers remove grease and scum, which minimizes membrane fouling. Installation of primary clarifiers may enlarge the plant footprint; primary clarifiers will also generate odors, which may have to be controlled.

4.4 Peak Flow Management

It is important to consider management of flow fluctuations caused by both diurnal variation in flow and storm events for end-of-pipe facilities. Depending on the size, age, and condition of a wastewater treatment plant (WWTP), the design peaking factor of the collection system typically varies between 2 and 4 times the average design flow and can be even higher for systems with significant infiltration and inflow (I/I). Membrane bioreactor facility designers have three options for handling peak flows: (1) provide sufficient membrane surface area to treat peak flow conditions through the membranes; (2) provide equalization volume upstream of the membranes to attenuate peak flows; and (3) provide a hybrid design where the MBR process operates in parallel with another unit operation and the treated effluent is blended for discharge. The design approaches may be combined to provide the most effective solution for a particular plant.

4.4.1 *Peak Flow Treatment with Membranes*

Because membrane sizing is hydraulically driven, alternatives to increasing the membrane surface area may be considered if the peak flow is more than twice the average flow (the typical economical upper limit of peak flow for membranes in most MBRs is approximately 2 to 3 times the average flowrate). Designing membranes to accommodate higher peak flows typically results in fluxes at the average flow condition that are below the optimized point, which, therefore, may increase equipment and O&M costs. In addition to cost, there are operational benefits associated with a constant and reasonable flux, including the opportunity to operate at optimized air scour rates. The combination of reduced membrane surface area and a lower air scour rate can result in significant reductions in energy use, as discussed in Sections 6.0 and 9.0.

4.4.2 *Equalization*

There are two options for equalization: external and internal. External equalization consists of separate tankage ahead of the biological process tankage, and can be used intermittently offline to store storm flows or inline to attenuate daily flows and peak-day flows. Internal equalization involves providing sufficient sidewater depth in the biological process tankage to allow variations in liquid level. In most plants, the variation will be limited to approximately 0.5 to 1.0 m before the aeration blowers are affected, unless positive-displacement blowers, which may be less efficient, are used. Typically, internal equalization is best suited for attenuating the diurnal flow pattern because the variation in liquid level required to effectively manage storm flows tends to be significant, which could adversely affect process blower design. For some facilities, a combination of both external and internal flow equalization, as shown in Figure 4.1, provides a cost-effective solution, with the external equalization basin used for offline storage of storm flows and internal equalization used to handle the daily diurnal variations in flow.

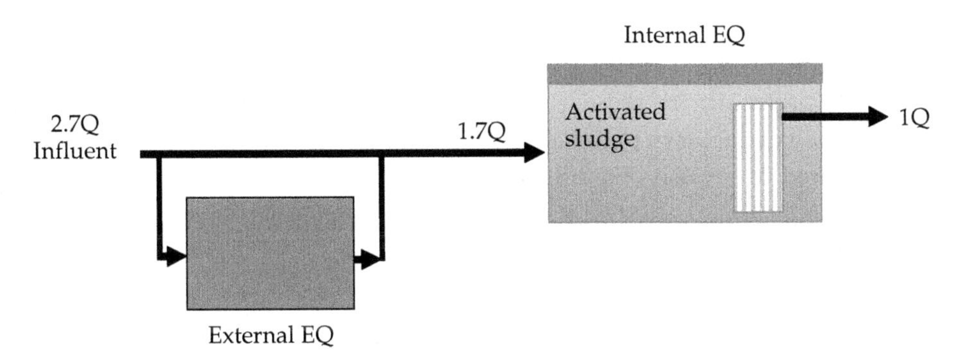

FIGURE 4.1 External and internal flow equalization. Note that Q is equal to annual average design flow.

4.4.3 *Hybrid Processes for Peak Flow Treatment*

Another option for treating peak flows in an MBR facility is the "hybrid process," which combines an MBR treatment train with a parallel unit operation. The concept is to design the MBR train for "base-load" conditions and treat flows or loads in excess of the base load in a parallel process that does not produce the same effluent quality; however, when the two effluent streams are blended, the resulting quality is sufficient. This strategy is commonly used when there is an existing parallel process, the peaking factor is high, or the peak occurs for an extended period of time. One example of a hybrid process is to include a chemically enhanced primary clarification train in parallel with an MBR. During peak flow events, the volume that exceeds the MBR design capacity is diverted to the chemically enhanced primary clarification process and the treated effluent from the two treatment trains is blended to meet treatment objectives. A second example is designing an MBR process train in parallel with a CAS process train. For more information on hybrid system design, refer to *Design of Municipal Wastewater Treatment Plants* (WEF et al., 2009).

4.5 Small Plant Considerations

For small plants (less than 0.3 ML/d), pretreatment systems are less mechanically intensive to accommodate remote operation or limited staffing. Screening devices such as trash traps or manually controlled screens may be preferable to automatic screen systems, although a fine-screen size opening of 1 to 3 mm is still required. Screenings from small systems, which serve as scalping plants, can be returned to the main sewer line, which eliminates the need for washing and compacting the screenings if compatible with the downstream treatment system.

4.6 Industrial Applications

The type of pretreatment required for industrial flows is highly specific to the industry. The key factor is removal of FOG, which can adversely affect the membranes as noted previously. Equalization is also an important consideration, especially for systems that receive multiple streams of industrial waste with a variety of characteristics.

5.0 BIOLOGICAL PROCESS

This section describes design considerations for a biological treatment process that are unique to MBR plant design. Chapter 3 contains further details on the fundamental aspects of biological wastewater treatment processes.

5.1 Configuration of Biological Reactor

The biological process design for an MBR is the same as the process design for a CAS plant, except that MBRs typically operate at higher MLSS concentrations and shorter hydraulic retention times (HRTs). For a given biomass inventory, an MBR can be operated at higher MLSS and with a smaller aeration basin volume and plant footprint. Biological nutrient removal (BNR) configurations for MBRs are similar to those for CAS systems. Process selection and configurations are dictated by treatment objectives. To achieve a stable biological process that minimizes fouling and yields good filterability, MBR systems are typically designed to nitrify fully (Trussell et al., 2007). Total nitrogen and/or phosphorus control requires a more complicated design. Denitrification requires an anoxic zone and a nitrified recycle stream. To remove phosphorus, MBR systems have typically included chemical addition, which reduces effluent phosphorus concentrations to low levels; however, as SRTs have reduced over time, as noted in subsequent sections, there is an increased evaluation and use of biological phosphorus removal. Because MBRs consistently achieve effluent TSS concentrations lower than 1 mg/L, any nitrogen and phosphorus associated with the solids are removed as well, which results in better effluent quality. Chapter 3 contains a more detailed discussion on process schematics for various BNR configurations.

In an MBR process configured for high-rate nitrogen removal (e.g., four-stage or five-stage BNR process), the membrane tank serves as the reaeration zone. Because the membrane packing densities differ among membrane configurations and because the volume of the membrane tank must be selected to accommodate the required membrane area, the reaeration zone is typically larger than what is needed for the reaeration process. The longer HRT in the membrane tank allows for additional endogenous decay and nitrification. Because some nitrates will be formed in the membrane tank, virtually complete denitrification is necessary in the second anoxic zone to achieve low nitrate concentrations in the final MBR effluent.

Oxygen transfer also influences basin sizing. With the capability to operate at higher MLSS concentration, sufficient biological mass can be maintained for complete treatment with a short HRT. However, the associated oxygen demand may be beyond the capability of conventional diffused aeration systems, hence requiring a larger basin to ensure the ability to meet oxygen demand.

5.2 Solids and Hydraulic Retention Time

A minimum HRT of raw wastewater is required to allow adsorption and synthesis of constituents before they are exposed to the membranes. Experience has indicated that a minimum HRT of 3 hours is required to allow influent colloidal matter to be adsorbed into flocs before reaching the membranes; however, this is a function of a number of variables, including wastewater characteristics and, therefore, the designer should consider the minimum allowable HRT for each specific application. In addition, the high RAS rate (typically 4 to 5 times the forward flow) results in a shorter HRT in all reactors than the HRT in conventional processes, which requires attention to kinetics and reactor design to avoid short-circuiting.

Whereas the earliest commercially applied MBRs were designed for SRTs longer than 25 days, there has been a general trend toward designing at reduced SRTs that are consistent with CAS plants with similar biological process configurations and treatment objectives (Daigger et al., 2010). As with a CAS plant, selection of a design SRT is a function of numerous parameters, including wastewater characteristics, treatment objectives, biological process configuration, and environmental conditions. Current practice is to design for an aerobic SRT that is sufficient to provide complete nitrification as there is evidence that a fully nitrified mixed liquor minimizes fouling potential (e.g., from extracellular polymeric substances [EPS]) (Trussell et al., 2007).

5.3 Mixed Liquor Suspended Solids Concentration

Operating at higher MLSS concentrations allows the use of smaller biological reactor volumes and, hence, smaller plant footprints compared to CAS systems. Most manufacturers recommend MLSS concentrations between 8000 and 12 000 mg/L in membrane tanks, with occasional operation up to 15 000 mg/L, which practically limits MLSS concentrations in the biological process to 8000 to 10 000 mg/L. The biological process in an MBR can be operated at lower MLSS concentrations by either (1) also operating at reduced MLSS concentrations in the membrane tanks or (2) reducing the RAS rate from the membrane tanks to the biological tanks.

5.4 Return Activated Sludge Recirculation

Membrane bioreactors require higher RAS rates than CAS processes, typically 4 to 5 Q vs 0.5 to 1 Q for conventional processes, to redistribute solids in the biological system and to avoid accumulation of solids in the membrane tanks. The RAS stream from the membrane tanks to biological basins can also serve as nitrate recycle; however, if the residual dissolved oxygen concentration in the membrane tanks is elevated, excess oxygen could be returned to the anoxic or anaerobic zone, thereby inhibiting denitrification and/or biological phosphorus removal. To avoid this problem, a separate mixed liquor recycle stream can be added or the dissolved oxygen concentration in RAS can be reduced using a deoxygenation tank before it is delivered to the anoxic or anaerobic zones. With two recycle streams, the mixed liquor recycle stream returns MLSS from the aerated zone to the anoxic zone and RAS returns MLSS from the membrane tank to the aerated zone. A third recycle stream may be required for BNR plants. Using two recycle streams results in greater operating flexibility, but increases the number of pumps and related energy costs. Creative configurations involving low-head pumps can limit such costs. Deoxygenating the RAS stream will require more basin volume, which can result in higher capital costs. Reducing the oxygen concentration in RAS does not guarantee sufficient reduction in the concentration of nitrates and, hence, delivering RAS to the head of the plant could inhibit biological phosphorus removal for BNR plants.

5.5 Activated Sludge Characteristics

Unlike a CAS process, sludge settleability is not a concern with MBR systems; however, filamentous bulking still needs to be controlled because of the effects of filaments on membrane fouling (Meng et al., 2006). What is important in the case of MBRs is sludge filterability, which is a measure of sludge quality and an indicator of membrane fouling and cleaning requirements. Sludge filterability may be affected by the fines and colloidal material in the mixed liquor, elevated concentrations of EPS, and floc size and characteristics. Chapter 7 presents tools and techniques to measure and quantify activated sludge characteristics in an MBR plant to support plant operations.

5.6 Oxygen Transfer

The design of a process aeration system for an MBR is the same as that for a CAS process, with a few exceptions. Because the alpha factor and the MLSS concentration are inversely proportional, a higher MLSS concentration in an MBR requires more aeration to deliver the same amount of oxygen (WERF, 2004). It is also not uncommon for the oxygen uptake rate to control the size of an aeration basin because of short HRTs that are typical for an MBR process.

The main function of membrane air scour is to prevent buildup of solids on the membrane surface; however, depending on the process configuration, a portion of the membrane air scour can be used toward biological process requirements, commonly referred to as *oxygen credit*. There are two ways that the oxygen credit can be applied: (1) oxygen consumed within the aerated membrane tank volume, and (2) oxygen supplied within the RAS stream. For systems that have no low-to-moderate total nitrogen removal requirements, both sources of oxygen credit can be considered. However, with increasing total nitrogen removal requirements, particularly in systems that include a post-anoxic zone, the nitrification process must be completed in the nitrification basin and, therefore, the oxygen credit for the oxygen consumed in the membrane tanks does not apply.

6.0 MEMBRANE SEPARATION PROCESS

6.1 Membrane System Configuration

In an MBR system, membranes replace the solids separation function of secondary clarifiers and sand filters in a CAS system. The membrane filtration system in an MBR typically consists of large membrane subunits arranged into membrane tanks or trains. The trains contain large membrane subunits that house a finite number of small membrane subunits, each of which has a defined surface area. It is the total surface area of the membranes installed in a plant that determines the plant capacity. That is, the more flow required, the more small membrane subunits are required.

The modular design of an MBR allows flexibility in the design of plant expansions. In many facilities, future capacity is built into the membrane trains by leaving space for future addition of large or small membrane subunits. The design flow of each train can be increased by adding membranes, thereby increasing the membrane surface area per train. Another common practice is to include spare space in the membrane tanks for the future addition of large or small membrane subunits as a design safety factor. The selection of the number and configuration of membrane trains should also consider reduced plant flows at initial operation if this is applicable to the plant.

Each membrane train has dedicated permeate piping and, if applicable, pumping equipment. The hydraulic capacity of each membrane train is determined by the maximum hydraulic capacity through the membrane system divided by the desired number of membrane trains in service for all flow conditions, including any membrane redundancy restrictions. The size of the equipment dedicated to operating a membrane train is determined by the membrane surface area in the train. Phasing-in capacity is typically designed in one of two ways. The first is to divide the planned future capacity within the initially installed trains by leaving additional space for adding membranes and ensuring that the installed equipment is hydraulically sized to handle future flows. Although this approach requires additional expenditure at the initial phase of design and construction, it minimizes the cost and time required to expand plant capacity and, as such, is often considered when future flow requirements are expected within a relatively time period. Providing additional surface area within the train typically limits future expansion to the same membrane filtration product. The second phasing design option is to plan for the addition of future membrane trains and equipment alongside the initial phase by either building all infrastructure, such as membrane tanks, pads, and piping, or simply leaving the footprint area free with no additional equipment installed. This approach defers costs to the future phase and is typically preferred when the future capacity requirement is not well defined or is not expected for some time.

An MBR can be either of greenfield construction or it can be retrofitted to an existing WWTP. Existing tanks can be used to accommodate biological processes and for membrane installation. Other options include a combination of repurposing existing tanks and building new tanks for biological and membrane processes.

6.2 Flux Selection

Flux selection is a balance between cost and risk. A lower design flux requires a larger membrane surface area and, therefore, higher capital costs; however, a lower design flux results in lower risk, longer membrane life, and, typically, less maintenance. Conversely, a higher design flux requires less membrane area and lower capital costs, but higher risks, shorter membrane life, and potentially higher maintenance costs. Effects on life-cycle cost should be considered when evaluating the design flux for a plant in addition to demonstrated long-term performance for membrane products.

Design flux selection must consider the composition of wastewater and mixed liquor for each application and should always be determined considering the performance of the membranes within their transmembrane pressure (TMP) operating range. Minimum requirements for selecting a design flux include design flows and temperatures, MLSS concentration, and ranges for each.

Flux selection must consider all flow conditions for the plant including membrane redundancy requirements. If the peak flow exceeds two times the average flow, the peak flow condition will typically govern the membrane design and determine the required membrane surface area; therefore, the installed membrane surface area will exceed that which is required for the average flow condition. In this instance, equalization may be considered as a means to attenuate peak flows and, thus, minimize the membrane surface area required (see Section 4.4).

Use of flux-enhancing chemicals (e.g., polymers and/or coagulants) has been proposed to reduce the required membrane surface area and as a means of increasing the short-term capacity of the membrane system for plants with high peaking factors. Flux enhancers are thought to improve membrane performance by removing soluble microbial products and EPS (Yoon et al., 2007). Published literature on this subject reveals mixed results, with some studies indicating positive results as long as overdosing is avoided (Koseoglu et al., 2008) while other studies indicate a potential for biological toxicity (Iversen et al., 2008). The potential benefits of flux-enhancing chemicals must be balanced against additional operating costs and complexity and, if used, polymer compatibility with membranes must be ensured.

6.3 Mixed Liquor Suspended Solids Concentration

Most manufacturers of immersed membrane systems recommend MLSS concentrations between 8000 and 12 000 mg/L in the membrane tanks, with occasional operation up to 15 000 mg/L. Some flat-sheet membranes extend the upper concentration limit to 15 000 to 20 000 mg/L. Higher concentrations are possible, but may require lower flux because of an increase in filtration resistance and higher operating costs for additional cleaning and reduced oxygen-transfer efficiency (e.g., membrane sludge thickening applications). Membrane bioreactor membranes can also be operated at lower MLSS concentrations; however, for membranes with larger pore sizes, there is a lower MLSS concentration limit because of the need to develop a minimum cake layer on the membrane surface to prevent pore fouling.

6.4 Return Activated Sludge Recirculation

To avoid accumulation of solids in the membrane tank, the MBR process includes an RAS stream (sometimes referred to as *mixed liquor recirculation*) where concentrated solids from the membrane tank are returned to the biological process. The RAS rate (*R*) is selected by comparing the design MLSS concentration in the biological process and the target MLSS concentration in the membrane tank. A mass balance around the membrane tank (Figure 4.2) results in the following equation that defines the

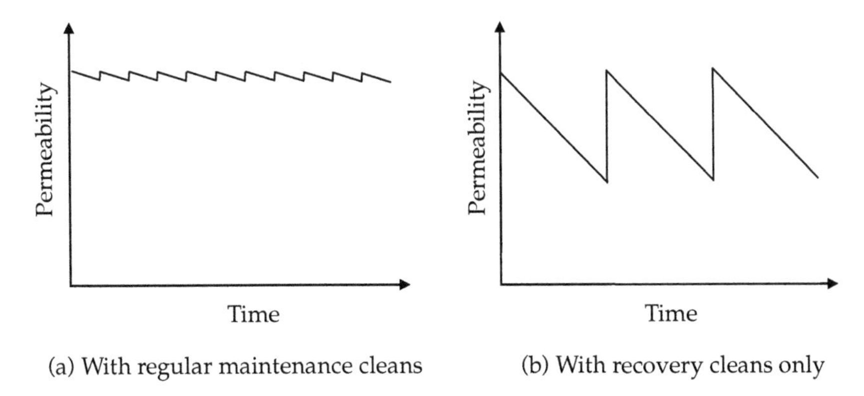

(a) With regular maintenance cleans (b) With recovery cleans only

FIGURE 4.2 Membrane bioreactor membrane tank mass balance.

relationship between the MLSS concentration in the biological process and the membrane tank:

$$\text{MLSS}_{\text{membrane tank}} = (R + 1)/R \times \text{MLSS}_{\text{bioreactor}} \tag{4.1}$$

Figure 4.3 shows the effect of varying RAS rates (R) on the MLSS concentration in the membrane tank with a given MLSS concentration of 8000 mg/L in the bioreactor. The recirculation rate for an MBR is typically designed at 3 to 5 times the forward flow because it results in a reasonable compromise between recirculation pumping energy and membrane tank MLSS concentration increase. For example, if the design MLSS concentration in the biological process is 8000 mg/L and the target MLSS concentration in the membrane tank is 10 000 mg/L, the required recirculation rate is 4 times the forward flow. This relationship is illustrated in Figure 4.4. Facilities with a wide range of operating flow conditions may be designed to allow the recirculation flow to be turned down so that it paces with actual plant flow to maintain a target MLSS concentration while reducing pumping costs.

6.5 Permeation

Permeation, or filtration, is the process of withdrawing clean water from the membrane system by applying a driving force to overcome all losses in the system, including TMP, friction losses, and a potential static lift or pressurized discharge. Whereas

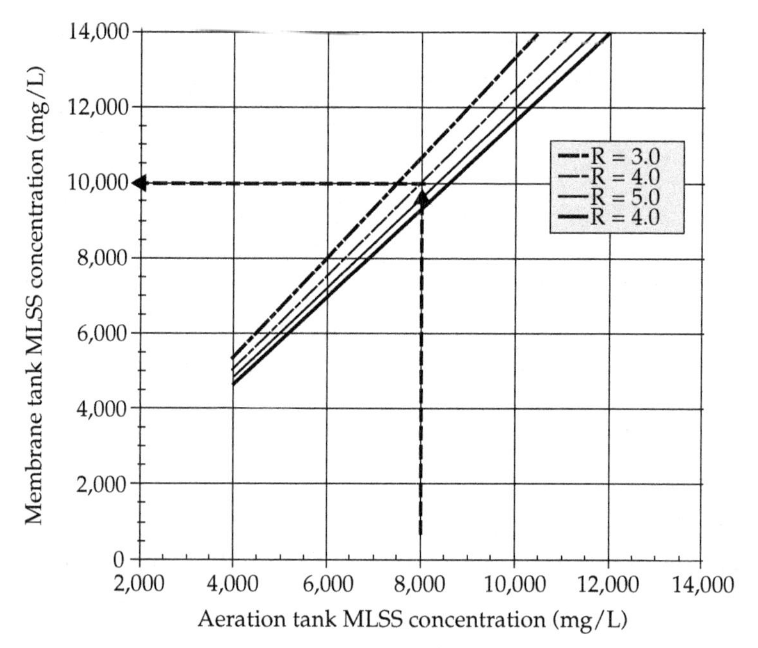

Figure 4.3 Membrane tank MLSS concentration as a function of RAS sludge rate (R) for a bioreactor MLSS concentration of 8000 mg/L.

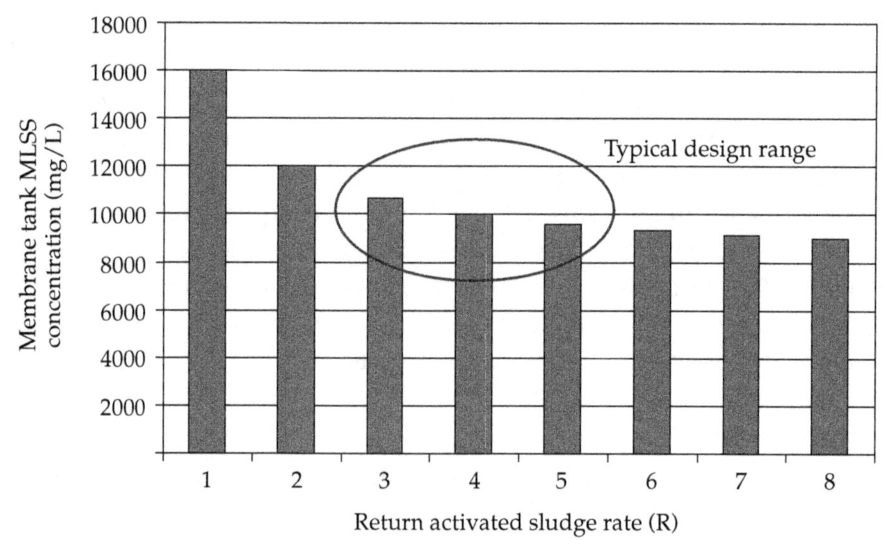

Figure 4.4 Relationship between aeration tank MLSS concentration and membrane tank MLSS concentration for varying RAS rates (R).

the TMP operating range is membrane-product specific and generally ranges from 3 to 21 kPa for flat-sheet membranes and 3 to 55 kPa for hollow-fiber membranes, other losses are site-specific and need to be considered regardless of membrane type.

Depending on the hydraulic gradeline of the plant, for immersed systems the necessary vacuum can be applied by pumping, gravity (siphon principle), or a combination of both (pump-assisted gravity). If a siphon system is used, a flow control valve replaces the pump for control of the permeate flow. The pump-assisted gravity design includes a pump to draw water through the membranes during peak flows and at elevated TMPs when the available head is insufficient.

6.6 Fouling Control

Fouling control is important to maintain the production capacity of the membrane system. All membrane technologies and configurations use a combination of physical and chemical cleaning. The basic cleaning methods are summarized in the following sections.

6.6.1 *Liquid-Velocity-Induced Shear*

For low-flow and specialized industrial applications, external tubular membrane systems may be a viable option. These membrane systems use liquid-velocity-induced shear at the internal surface of the membrane to control membrane fouling. For these applications, membrane air scouring and production cycling are typically not required.

6.6.2 *Membrane Air Scouring*

For all immersed MBRs and some external MBRs, air scouring is accomplished by medium- or coarse-bubble aeration to remove solids from the membrane surface. Aeration systems differ among vendors, and can be an integral part of a large membrane subunit or a separate assembly within the membrane tank.

Air scour can represent one of the largest energy uses in the MBR process. Many membrane manufacturers have recently reduced energy requirements of air scouring, and further improvement is anticipated. This has been one of the factors making MBR processes more competitive with CAS processes. Several membrane manufacturers have reduced the energy consumption of their systems by either (1) intermittent air scouring, which can be accomplished within large membrane subunits, membrane subtrains, or between membrane trains; (2) lower air scour flowrates at lower flux; or (3) a combination of both. Typical current rates of air scour are 0.1 to 0.6 Nm^3/h per square meter of membrane area.

As noted in Section 5.6, depending on the configuration of the biological and membrane separation processes, membrane scour aeration can provide a portion of the biological process air, which is referred to as an *oxygen credit*.

6.6.3 *Production Cycling—Relaxation and Backwashing*

All immersed and some external MBR systems operate with a production cycle, which consists of a permeation step followed by either a relaxation or backwashing step. The production cycle is a fully automated process and generally lasts 5 to 15 minutes. In relaxation mode, each train will permeate for 5 to 15 minutes and then undergo a period of relaxation with no permeate flow for 30 to 120 seconds. During relaxation, permeation stops, releasing the vacuum while aeration continues.

Whereas relaxation is the normal mode of operation, some membranes (typically hollow fiber) can operate in backwashing mode. In backwashing mode, flow through the membranes is not only stopped, but reversed through the membranes from the inside out for 30 to 60 seconds every 5 to 15 minutes. Backwashing water is permeate produced during the permeation step of the production cycle. This is accomplished by storing a volume of permeate in a backwash tank or by using an oversized common discharge header. Backwash flowrates, or fluxes, are specific for each type of membrane, but are typically 1 to 1.5 times the maximum instantaneous production permeate flux.

Backwash, or relax cycles, for all membrane trains can be either grouped together back to back at the beginning of the production cycle or staggered throughout the production cycle. Typically, only one membrane train is in backwash or relax mode at a time.

6.6.4 *Chemical Cleaning*

Regular chemical cleaning is a critical factor in maintaining the long-term performance of a membrane system. Cleaning strategies and methods vary among membrane types in terms of automation, frequency, and type and concentration of chemicals. There are two categories of chemical cleaning: maintenance cleaning and recovery cleaning. Using a combination of the two extends the life of membranes by maintaining their permeability and limiting their exposure to chemicals. Regular maintenance cleanings maintain a low operating TMP, allowing the system to respond to greater hydraulic loads and extending the interval between recovery cleanings (Figure 4.5).

The two primary cleaning agents used for MBR applications are sodium hypochlorite, for organic fouling, and an acid cleaner, typically citric or oxalic acid, for inorganic fouling. The wastewater application and system-operating condition will dictate which of the two is used and at what frequency. Each membrane type may have a limitation on chemical contact time over the life of the membrane.

6.6.4.1 Maintenance Cleaning Regular maintenance cleaning maintains membrane permeability between recovery cleanings. The maintenance cleaning is done in situ, is entirely automated, can be scheduled for off-peak hours of the day, and typically lasts between 0.5 and 2 hours. The membrane train is removed from service and the chemical is introduced from the inside to the outside of the membrane, either continuously or in pulses, while the membrane remains submerged in mixed liquor. Following a predetermined chemical contact time, the membrane train is returned to service and any residual cleaning solutions are returned to the biological process in RAS to be consumed or neutralized.

For systems that implement a "maintenance clean" with municipal wastewater, typically one to two maintenance cleans per train are performed per week. For industrial wastewater systems, the maintenance clean frequency is site-specific.

6.6.4.2 Recovery Cleaning Recovery cleaning is more intensive than maintenance cleaning, with a higher chemical concentration and a longer duration to recover

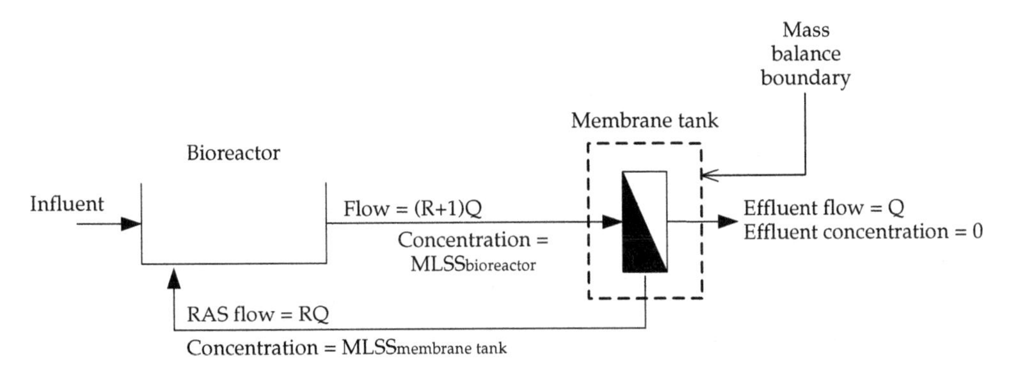

FIGURE 4.5 Typical permeability behavior with different cleaning regimes.

membrane performance. Recovery cleaning is completed using a soak solution made up of permeate produced by the plant and the selected cleaning chemical. Depending on the membrane type and plant size, recovery cleaning can be done in situ (i.e., within the membrane tank) or ex situ (i.e., in an external basin), and either with or without draining the mixed liquor from the membrane tank. Ex situ cleaning is typically used only at small plants with a small number of membranes. In all in situ cases, recovery cleaning can vary from fully manual to fully automated; moreover, the level of automation should be determined by balancing the capital and operating costs for each application. At a minimum, ex situ recovery cleans require some manual intervention to move the large membrane subunits; in addition, they are typically fully manual processes.

An in situ recovery cleaning involves isolating the membrane train from service and cleaning in one of the following two ways: (1) draining the mixed liquor and replacing all the contents of the membrane tank with a soak solution, or (2) retaining the mixed liquor in the membrane tank and backfilling the membranes with a soak solution. The soak duration lasts from 4 to 24 hours. For the first recovery cleaning method, the spent solution can be managed by adding a small portion of mixed liquor to the membrane tank to consume the remaining chemical, having chemical neutralization in place or in another basin, or draining the tank to be mixed with incoming wastewater to consume the chemical. For the second cleaning method, initial permeate is redirected either to another basin for chemical neutralization or to the head of the plant to be consumed by incoming wastewater until the chemical concentration in the permeate has reached acceptable levels for discharge. The ex situ cleaning method is similar, with the exception that individual large membrane subunits are removed from the membrane train and inserted in a small cleaning tank for soaking.

6.6.5 *Mechanical Cleaning*

Mechanical cleaning consists of cleaning the membranes by hand or by machine (e.g., low-pressure water spraying or other methods) to remove foulants or dewatered solids. In MBR systems, localized dewatering may occur whereby "dried" or more concentrated solids accumulate in and around membrane equipment (Codianne, 2008). Under normal operation, this type of fouling is not expected, but may occur as a result of pretreatment being bypassed, loss of membrane aeration, or other major mechanical failure.

Localized dewatering can lead to the formation of refractory concentrated cake that is difficult or impractical to remove by common cleaning procedures and often must be manually removed. The most common cleaning method is by low-pressure water spray; however, some membrane vendors have developed purpose-built cleaning equipment specific to their product (Codianne, 2008).

At least one membrane vendor has proposed combining the use of plastic media with membrane aeration in the filtration tank as a means of continuous mechanical cleaning.

6.7 Ancillary Equipment

In addition to permeate extraction and cleaning equipment, the components outlined in the following sections are commonly included in an MBR membrane filtration system.

6.7.1 *Air-Extraction System*

An air-extraction system is used to remove accumulated air from permeate suction piping to avoid air locking of pumps or other equipment. Air accumulation can result from entrained air coming out of solution under vacuum or from minor leaks in piping, typically after extended periods of system downtime. Types of air-extraction systems include ejectors, vacuum pumps, and self-priming permeate pumps. This equipment is controlled by the membrane filtration system, programmable logic controller (PLC), and is operated continuously or initiated by the production step or at operator-defined intervals.

6.7.2 *Membrane-Integrity Monitoring*

The industry-accepted practice for monitoring membrane integrity in an MBR system is continuous online permeate turbidity measurement. An increase in permeate turbidity indicates either a membrane breach or a leak in system piping that has allowed mixed liquor or air to enter the permeate stream. The operator can be notified of this by the PLC and can further diagnose the problem by isolating membrane trains, large membrane subunits, and small membrane subunits. Hollow-fiber membranes can also be pressurized with air to detect the location of the breach.

6.8 Instrumentation and Process Control

The membrane separation process system includes a variety of equipment and instrumentation controlled by a PLC. The PLC may be dedicated solely to membrane equipment or may include controls for other equipment. The PLC functions include controlling operating procedures (e.g., automated cleaning), alerting or alarming certain conditions, turning equipment on and off based on demand, and trending operating information such as flow, permeate turbidity, and TMP. It may also communicate with a plant supervisory control and data acquisition system to provide additional trending functionality and integrate the membrane system operation with the remainder of the plant. Because an MBR system requires a functional PLC, attention to PLC redundancy requirements is important.

7.0 POST-TREATMENT

Membrane bioreactor post-treatment considerations are similar to CAS treatment systems, except that MBR effluent offers some advantages for post-treatment design. Although membrane filters in an MBR process are not considered disinfection devices, an MBR system does achieve bacteriological quality that generally meets direct-discharge requirements. In addition to low bacteriological counts, the retention of particulate and colloidal material by membranes produces a high-quality effluent that has high UV-transmissivity values (typically greater than 75%), which reduces operating costs for UV disinfection, and low chlorine demand, which may reduce contact-time requirements for disinfection. The treated effluent from an MBR process is also well suited for further advanced treatment by reverse osmosis or advanced oxidation processes.

When combining downstream post-treatment systems with an MBR, an important consideration is compatibility of the control systems. Generally, post-treatment is designed as a continuous process and MBRs will operate on a "semi-discontinuous" basis, with additional flow fluctuations beyond diurnal- and seasonal-flow variations. These additional fluctuations are caused by the requirement of periodic relaxation or backwashing of the membranes, which regularly interrupts production from a portion of the membrane system, and filling of the backpulse tank if included in the design. This effect is more apparent in smaller systems with few trains or at lower flows, and will result in short periods of reduced or even zero flow. The post-treatment control system must recognize these events and continue to operate, requiring coordination between the control systems of the MBR and the disinfection process.

8.0 RESIDUALS TREATMENT

8.1 Screenings

As discussed in Section 4.2, fine-screening equipment can generate a substantial amount of screenings. Section 4.2 presents additional details and design considerations for collecting and managing this residual.

8.2 Waste Activated Sludge

The production of waste activated sludge (WAS) in an MBR is similar to a CAS process designed with equivalent SRTs. Although retention of more fine and colloidal materials in the membrane separation process and smaller floc size may result in poorer mixed liquor settleability than in conventional systems, WAS from an MBR can generally be dewatered to concentrations similar to those of WAS in a CAS plant (Schwarz et al., 2006). There is contradictory evidence in the literature on the relative dewaterability of WAS from an MBR vs a CAS process. Currently, however, designers typically follow the same practice for dewatering process and equipment design for an MBR as they would for a CAS plant. The volume of WAS produced in an MBR can be reduced by extracting WAS from RAS from the membrane tank where the MLSS concentration is highest.

The location where solids processing sidestreams are returned to the liquid train in an MBR process is important because these streams may contain polymers that could be fouling in nature. It is common practice to return these streams to the head of the plant so that any residual polymer can be adsorbed in the biological process before it reaches the membrane tanks. Solids-processing sidestreams can also be high in nutrients; this phenomenon needs to be considered in the design and operation of systems that have nutrient-removal requirements.

8.3 Foam and Scum

As with any biological system, foam and scum may be generated on the mixed liquor surface. In municipal systems, there generally will only be a relatively stable layer at the tank surface, which may be an aesthetic nuisance, but not an operational problem. Industrial systems often do not require any additional components for foam and scum management, except for additional freeboard for retention and, potentially, a sprayer system. Similar to a CAS system, there is the potential for thick foam and scum layers to develop, which can cause instrument failure or tank overflows.

As in conventional systems, short-term solutions that are effective in foam management include surface chlorination or the use of chemical solutions such as antifoams and defoamers. It is important to note that membrane compatibility with antifoams or defoamers should be confirmed with the membrane supplier. Surface wasting has been shown to be an effective long-term solution to deal with foam and scum. If surface wasting is being provided, a review of mixed liquor flows through the entire system is required to ensure this is effective for eliminating the potential foam and scum traps that would prevent the material from reaching the wasting point. A combination of overflow weirs, partially submerged gates, and sprayers can be used to enable foam collection.

8.4 Spent Cleaning Solutions

A waste stream particular to MBRs is spent membrane cleaning solutions. Generally, spent cleaning solutions consist of an organic membrane cleaner that is an oxidant, commonly sodium hypochlorite, and an inorganic membrane cleaner that consists of citric acid or oxalic acid and may also contain some stronger acids such as hydrochloric, sulfuric, or nitric acid.

A common method of managing spent cleaning solutions is to sacrifice a small portion of the bacteria in the mixed liquor to consume the remaining cleaning chemicals before a membrane train is returned to service. This method is typically used at medium-to-large plants, and may not be suitable for small plants. Consumption of the chemical solution with mixed liquor is generally completed in one of two ways: (1) if the membranes are soaked in the solution, the membrane basin feed valve, or gate, is used to introduce a volume of mixed liquor to be mixed with the chemical solution, or, (2) if the cleaning solution is contained within the membranes submerged in mixed liquor, clean permeate is used to displace the cleaning solution into the mixed liquor where it will be consumed.

If the use of mixed liquor is impossible or undesirable, chemical neutralization systems either within the membrane tank or via transfer of the chemical solution to another basin can be used. Typically, a combination of caustic and sodium metabisulphite are used to dechlorinate sodium hypochlorite solutions, and caustic is used to neutralize acid based cleaners. Alternatively, spent cleaning solutions can be drained to the plant feed system, where raw wastewater can consume the chemical residuals, or they can be discharged to the sewer and conveyed to a downstream wastewater plant for treatment. Materials of construction of components that come in contact with initial and spent cleaning solutions need to be considered.

9.0 ENERGY OPTIMIZATION

Relatively high energy consumption is one of the disadvantages of MBR technology. Improvements in membrane products and plant design have resulted in a significant reduction in energy requirements of MBR plants. This section highlights design and operational strategies to reduce overall energy consumption of an MBR.

9.1 Design Elements to Reduce Energy

There are several focus points that provide the opportunity for cost-effective and energy-efficient design of an MBR plant that balances capital and operating costs. These points include the use of primary clarification ahead of the MBR, flow equalization, adjusting the balance of solids between the aeration and membrane basins, and pump configuration. The degree of savings both in energy and system costs depends on plant size and local economics.

As noted previously, equalization provides an opportunity to reduce both the membrane surface area required for peak flows and the flux variation, which can occur from diurnal flows and/or peak flows. Limiting the flux to a reasonable value also enables most manufacturers to operate at a reduced air scour rate, as discussed in Section 6.6.2. The combination of smaller membrane surface area and operating with a lower air scour rate can significantly reduce the energy requirements.

9.1.1 *Balance of Solids*

Traditionally, MBR systems have been designed to operate at similar MLSS concentrations in both biological process basins and membrane tanks. The result is a high solids recirculation rate of 4 to 5 times the influent flow. Although it is not feasible in all MBR designs, under certain circumstances (e.g., with the use of primary clarifiers) there is an opportunity to operate with lower MLSS concentrations and, hence, less mass in the aeration basins yet maintain a higher MLSS concentration in the membrane tanks. This mode of operation could reduce the recirculation rate up to 50%. Energy reduction resulting from this configuration is two-fold: (1) reduction in pumping, and (2) a potential increase in alpha factor, which improves oxygen-transfer efficiency.

9.1.2 *Pump Configurations*

As discussed in Sections 5.4 and 6.4, the three key pumping requirements for an MBR are RAS, nutrient recycle, and permeate. In a traditional MBR design, which provides full BNR, there is the potential of up to 11 Q of pumping (5 Q for RAS, 4 Q for denitrification, 1 Q for biological phosphorus removal, and 1 Q for permeate pumping). As described previously, operating with a difference in MLSS concentrations between the membrane basins and aeration basins could reduce RAS flow by 50%. Use of a deoxygenation zone for RAS would allow return of the solids to the anoxic zone without detrimental effects from residual dissolved oxygen and, hence, facilitate denitrification. This would reduce the additional mixed liquor recycle for denitrification by 50%, yet still yield a total recycle ratio of 4 Q for denitrification. Using gravity flow from the membranes rather than permeate pumps would reduce pumping by an additional 1 Q. However, this configuration would require sufficient water depth above the membranes to offset headloss associated with flux variation, fouling of the membranes, and any downstream processes prior to discharge. If pumping is required to reach the discharge point, permeate pumps may be the most cost-effective choice. Chapter 5, Section 2.3, contains more details on membrane permeation.

9.2 Operational Elements to Reduce Energy

In addition to the design elements discussed previously, there are various operational elements that influence overall energy efficiency of the MBR design. Currently, the largest item of energy cost is the cost of aeration, both for the biology and maintenance of the membranes, as illustrated in Figure 4.6. Hence, reducing aeration has the potential to reduce the overall energy requirements substantially.

9.2.1 *Membrane Air Scour*

As discussed in Section 6.6.2, a key factor in the performance of the MBR process is providing membrane air scour. To reduce energy use, some membrane manufacturers have developed strategies to reduce air scour requirements (see Section 6.6.2).

9.2.2 *Optimize Membranes in Service*

Matching the number of membrane trains in service with plant flow is an operating strategy that can reduce energy use because the membranes that are not in service do not require the same degree of air scour as those in service. Consequently, taking membrane tanks out of service during the times of the day when flow is low provides an opportunity to reduce air scour requirements during these periods. This strategy

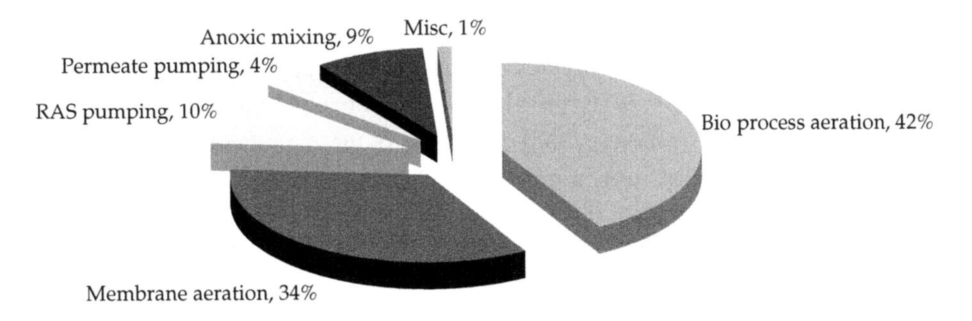

FIGURE 4.6 Energy requirements for an MBR (Hribljan, 2007).

requires cycling membrane basins in and out of service to maintain the membranes available for service and to maintain the activity of the biology in the tanks. Although manufacturer-specific, the air scour can be reduced to minutes out of a 30-to-60-minute period. This mode of operation may also enhance membrane performance because of a more consistent flux. There is evidence that operating at a constant and optimum flux results in decreased membrane fouling and increased membrane life. Varying the number of basins online is primarily an opportunity for plants where equalization is not provided.

9.2.3 *Optimize Dissolved Oxygen within the Biological Process*

In all WWTPs that use aerobic treatment, biological aeration demand is a significant contributor to the plant's energy use. There are two opportunities to reduce total aeration demand in biological aeration basins of an MBR: (1) operate at the minimum dissolved oxygen required to achieve complete treatment, and (2) return the solids from the membrane tank to the aerobic portion of the biological basins to use the elevated dissolved oxygen in the RAS stream from the membrane tanks. Historically, the aerobic portion of biological basins has been operated with a target dissolved oxygen concentration of 2 mg/L to consistently achieve performance goals and to minimize the potential for filamentous growth. By using membranes instead of gravity settling systems for solids separation, the adverse effect of filaments on settling is reduced; however, research has shown that filaments can cause fouling of membranes (see Chapter 2, Section 4.3). Consequently, aerobic basins could be operated with a residual dissolved oxygen concentration of 1 mg/L, or potentially less, to reduce aeration demands. However, it is critical to have sufficient SRT and HRT to achieve the required performance, especially with a reduced concentration of dissolved oxygen. With respect to the solids recycle line, returning the solids from the membrane tank to the anoxic zone (if it is part of the biological basins) could be detrimental to denitrification because of the elevated dissolved oxygen concentration. Depending on the membrane manufacturer, membrane aeration rate, and process type, the dissolved oxygen concentration in the membrane tank can vary between 2 and 6 mg/L. Returning solids to the aerobic zone provides an opportunity to use dissolved oxygen to offset a portion of aeration demand.

9.3 Equipment Elements to Reduce Energy

As with conventional suspended growth activated sludge processes, equipment selection can have a significant effect on reducing energy costs. Operation of diffused aeration systems, biological process blowers, membrane air scour blowers, and mixers typically constitutes the largest energy use in an MBR plant. Chapter 5 contains more information on MBR facility and equipment design.

10.0 PROCESS SIMULATION FOR MEMBRANE SYSTEMS

Mechanistic models such as the IWA ASM1 (IWA Task Group on Mathematical Modelling, 2000) and its later variants have provided valuable insight into the activated sludge process and enabled engineers to better understand the significance of different parameters used to describe activated sludge system behavior. Whereas decades of experience are built into current models, MBR models are still in the development stage. Although bio-transformations in an MBR are the same as in CAS, the models do not yet describe how biological and membrane processes interact. A few simulation models available in the market account for fouling and cleaning cycles, but there are not enough full-scale data to calibrate them. More research is required to address the challenges faced with modeling MBRs and to answer outstanding questions about the effect of increased MLSS concentrations on biological kinetics and the relationship between the biological process and hydraulic performance of the membrane system.

Process simulation models are powerful tools for designing and evaluating WWTPs. In past years, these models have been successfully applied to the design and optimization of MBR facilities (Daigger et al., 2010). Moreover, the models can be used to evaluate different operating mixed liquor concentrations, different recycle flow configurations, effects of different air scour regimes on membrane tank dissolved oxygen, and the effect of membrane tank dissolved oxygen on the upstream process. Modeling an MBR is particularly beneficial when designing for nutrient removal. Elevated dissolved oxygen concentrations in recycle streams can contaminate upstream anoxic and anaerobic zones and can impair performance of processes with limited influent carbon. Simulating an MBR can also help determine the effect of reduced HRTs in secondary anoxic zones caused by elevated RAS flowrates.

11.0 REFERENCES

Codianne, B. (2008) Enviroquip MBR Systems: The Latest Innovations. *Workshop at the 81st Annual Water Environment Federation Technical Exhibition and Conference;* Chicago, Illinois, Oct 18–22; Water Environment Federation: Alexandria, Virginia.

Côté, P.; Brink, D.; Adnan, A. (2006) Pretreatment Requirements for Membrane Bioreactors. *Proceedings of the 79th Annual Water Environment Federation Technical Exhibition and Conference* [CD-ROM]; Dallas, Texas, Oct 22–25; Water Environment Federation: Alexandria, Virginia.

Daigger, G. T.; Crawford, G. V.; Johnson, B. R. (2010) Pilot- and Full-Scale Assessment of the Nutrient Removal Capabilities of Membrane Bioreactors. *Water Environ. Res.*, **82** (9), 806.

Hribljan, M. J. (2007) *Large MBR Design and Residuals Handling* [Webcast]; June 12; Water Environment Federation: Alexandria, Virginia.

Iversen, V.; Mohaupt, J.; Drews, A.; Kraume, M.; Lesjean, B. (2008) Side Effects of Flux Enhancing Chemicals in Membrane Bioreactors (MBRs): Study on their Biological Toxicity and their Residual Fouling. *Water Sci. Technol.*, **57** (1), 117.

IWA Task Group on Mathematical Modelling for Design and Operation of Biological Wastewater Treatment (2000) *Activated Sludge Models ASM1, ASM2, ASM2d, and ASM3;* IWA Publishing: London, U.K.

Koseoglu, H.; Yigit, N.O.; Iversen, V.; Drews, A.; Kitis, M.; Lesjean, B.; Kraume, M. (2008) Effects of Several Different Flux Enhancing Chemicals on Filterability and Fouling Reduction of Membrane Bioreactor (MBR) Mixed Liquors. *J. Membrane Sci.*, **320**, 57.

Meng, F.; Zhang, H.; Yang, F.; Li, Y.; Xiao, J.; Zhang, X. (2006) Effect of Filamentous Bacteria on Membrane Fouling in Submerged Membrane Bioreactor. *J. Membrane Sci.*, **272**, 161.

Metcalf and Eddy, Inc. (2003) *Wastewater Engineering Treatment and Reuse,* 4th ed.; McGraw-Hill: New York.

Schwarz, A. O.; Rittmann, B. E.; Crawford, G. V.; Klein, A. M.; Daigger, G. T. (2006) Critical Review on the Effects of Mixed Liquor Suspended Solids on Membrane Bioreactor Operation. *Sep. Sci. Technol.,* 41, 1489.

Trussell, R. S.; Jang, N.; Merlo, R. P.; Kim, I. S.; Hermanowicz, S. W.; Jenkins, D. (2007) Why Are MBRs Commonly Designed to Nitrify? Because Changes in Organic Foulant Properties Impact Membrane Fouling. *Proceedings of the 80th Annual Water Environment Federation Technical Exhibition and Conference* [CD-ROM]; San Diego, California, Oct 13–17; Water Environment Federation: Alexandria, Virginia.

Van der Roest, F. F.; Lawrence, D. P.; Van Bentem, A. G. N. (2002) *Membrane Bioreactors for Municipal Wastewater Treatment;* STOWA Report; IWA Publishing: London, U.K.

Water Environment Federation; American Society of Civil Engineers; Environmental and Water Resources Institute (2009) *Design of Municipal Wastewater Treatment Plants,* 5th ed.; WEF Manual of Practice No. 8; ASCE Manual and Report on Engineering Practice No. 76; McGraw-Hill: New York.

Water Environment Research Foundation (2004) *WERF Project #00-CTS-8a: MBR Website Strategic Research Final Report;* Water Environment Research Foundation: Alexandria, Virginia.

Yoon, S-H.; Collins, J.; Bhasker, D.; Koppes, J.; Nishimori, K.; Qin, J.; Livingston, D. (2007) Five Year Experience with Membrane Performance Enhancers for Membrane Bioreactors: From Mechanistic Studies to Full-Scale Implementation. *Proceedings of the 80th Annual Water Environment Federation Technical Exhibition and Conference* [CD-ROM]; San Diego, California, Oct 13–17; Water Environment Federation: Alexandria, Virginia.

12.0 SUGGESTED READINGS

Barnard, J. L.; Phillips, H.; Sabherwal, B.; deBarbadillo, C. (2008) Driving Membrane Bio-Reactors to Limit of Technology. *Proceedings of the 81st Annual Water Environment Federation Technical Exhibition and Conference* [CD-ROM]; Chicago, Illinois, Oct 18–22; Water Environment Federation: Alexandria, Virginia.

Buer, T.; Cumin, J. (2010) MBR Module Design and Operation. *Desalination*, 250, 1073.

Enviroquip (2007) Flux Documentation; Enviroquip: Austin, Texas.

Itokawa, H.; Thiemig, C.; Pinnekamp, J. (2008) Design and Operating Experiences of Municipal MBRs in Europe. *Water Sci. Technol.*, **58** (12), 2319.

Judd, S. (2006) *The MBR Book: Principles and Applications of Membrane Bioreactors in Water and Wastewater Treatment*; Elsevier: Oxford, U.K.

Krause, S.; Zimmerman, B.; Siembida, B.; Cornel, P. (2009) Mechanical Cleaning Process for Enhanced MBR-Performance. *Proceedings of the 82nd Annual Water Environment Federation Technical Exhibition and Conference* [CD-ROM]; Orlando, Florida, Oct 11–14; Water Environment Federation: Alexandria, Virginia.

Thompson, D.; Vonghia, E. (2007) Design & Implementation of Large-Scale Membrane Bioreactors in North America. *Proceedings of the Aachen Membrane Kolloquium* [CD-ROM]; Aachen, Germany, Mar 28–29.

Wallis-Lage, C. L.; Hemken, B. E.; deBarbadillo, C.; Steichen, M. T. (2005) MBR Plants—Larger and More Complicated. *Proceedings of the 78th Annual Water Environment Federation Technical Exhibition and Conference* [CD-ROM]; Washington, D.C., Oct 29–Nov 2; Water Environment Federation: Alexandria, Virginia.

Wallis-Lage, C. L.; Hemken, B.; deBarbadillo, C.; Steichen, M. (2005) Shopping for an MBR: What's for Sale? *Water Environ. Technol.*, **17** (1), 31.

Wallis-Lage, C. L.; Levesque, S. D. (2009) Cost Effective & Energy Efficient MBR Systems. *Proceedings of Singapore International Water Week;* Singapore, June 22–26.

Chapter 5

Membrane Bioreactor Facility Design

(continued)

(continued)

1.0 GENERAL CONCEPTS FOR DESIGN OF MEMBRANE BIOREACTOR FACILITIES

1.1 Introduction

This chapter focuses on the design and configuration of membrane bioreactor (MBR) facilities and their significant components. Chapter 4 presents a discussion on process design and development of design criteria for MBR systems.

The purpose of this chapter is to discuss overall concepts, configuration issues, materials, and other considerations associated with design and construction of MBR treatment facilities. The evolution of this treatment approach from small and satellite treatment systems to an accepted approach for end-of-pipe treatment facilities requires detailed consideration of design and construction issues that are specific to MBR processes. Currently, submerged (immersed) membrane technology dominates in the municipal market and, therefore, will be the focus of this chapter.

The emphasis of this discussion is on design and configuration associated with the liquids treatment system. Solids-handling design issues unique to MBR design will be identified. *Design of Municipal Wastewater Treatment Plants* (WEF et al., 2009) contains a more detailed discussion on the design of solids handling facilities.

1.2 Membrane Bioreactor System Unit Process Configuration and Layout

This section presents a brief discussion of the unit processes associated with MBR systems together with options that should be considered by the designer when configuring and laying out a facility using an MBR system. Unit processes used in treatment facilities with an MBR system may consist of the following processes:

coarse and fine screening; grit removal; fats, oils, and grease (FOG) removal; flow equalization; primary clarification; biological treatment; membrane separation; and disinfection. A general treatment train layout is shown in Figure 1.1 of Chapter 1. Depending on the project requirements, the MBR treatment train can range from a minimum number of process steps (such as coarse screening, fine screening, biological treatment, membrane separation, and disinfection) to all of the unit processes shown in Figure 1.1. A configuration with a minimum number of process steps could correspond to the design of smaller MBR systems in areas where land may be at a premium or the owner desires an enclosed facility. As other process and design tradeoffs are identified, and in situations where footprint is not a constraining factor, treatment train configurations using grit and FOG removal, equalization, and primary clarification may be considered. More complicated treatment trains are typical for MBR systems treating larger average and peak flows (> ~4 to 5 mgd), where the membrane equipment cost becomes large enough to warrant consideration of design alternatives other than providing sufficient membrane area to treat peak flows. Table 5.1 lists some facility alternatives to be considered in the design of MBR treatment plants. As designers progress from preliminary to final design and more

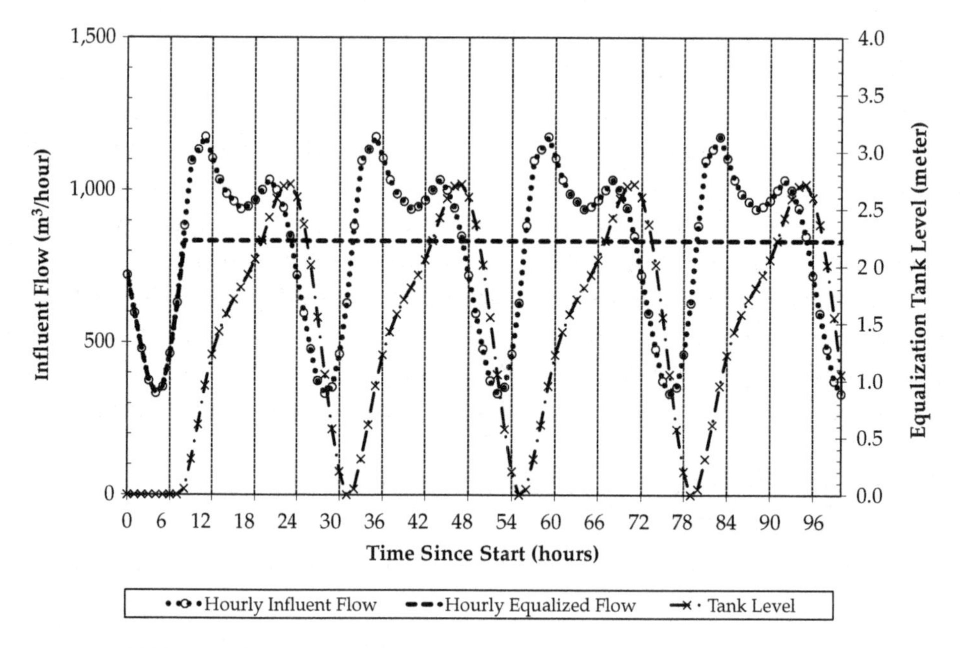

Figure 5.1 Effect of flow equalization on MBR flows.

TABLE 5.1 Some top-level design tradeoffs when configuring MBR treatment facilities.

	Unit process affected							
Design tradeoffs to consider	Coarse screening	Grit/FOG removal	Equalization	Primary clarification	Fine screening	Biological process	Membranes	Disinfection
Hybrid System: Dry weather MBR—wet-weather enhanced coagulation or CAS	X	X	X	X	X	X	X	X
Equalization or no equalization	X	X	X	X	X	X	X	X
Separate equalization tank or equalization in the biological tank			X		X	X		
Equalization sized to invert dry weather diurnal to reduce energy costs			X		X		X	X
Screening before or after influent pumping	X				X			
Is grit/FOG removal required		X	X		X			
Is grit/FOG treatment step required for membrane warranty		X					X	
Primary clarification or no primary clarification				X	X	X	X	
Fine screening before MBR or on a fraction of RAS					X		X	

site-specific information becomes available, design decisions will be made to help balance these competing objectives to provide a robust, flexible, and cost-effective MBR system design.

Peak flow management-related design tradeoffs including the use of a hybrid MBR-enhanced coagulation and CAS system or flow equalization affect the design of all unit processes, and require careful consideration during initial design (Table 5.1).

The remainder of Section 1.0 of this chapter addresses design issues and design decisions associated with unit processes upstream of the MBR facility and more general overall design considerations for MBR facilities. Section 2.0 discusses design issues associated with the biological treatment and membrane separation process of the treatment train. Section 3.0 covers equipment- and appurtenance-related design issues.

1.3 Peak Flow Management and Membrane Bioreactor Facility Design

Unlike conventional treatment facilities which, by design, can treat high-flow conditions and degrade in performance slowly, membrane-based treatment processes need to stay within their design range to ensure both performance and membrane longevity. Chapter 4 provides an overview of process solutions for peak flow management. The design of equalization basins is reviewed here. For hybrid design issues, refer to *Design of Municipal Wastewater Treatment Plants* (WEF et al., 2009) for conventional activated sludge treatment design.

1.3.1 Flow Equalization

There are several methods available for estimating the storage volume for flow equalization. For general procedures for sizing equalization facilities, refer to *Evaluation of Flow Equalization in Municipal Wastewater Treatment* (Ongerth, 1979). Because bioreactors in MBR facilities are designed with shorter hydraulic retention times (HRTs) than conventional activated sludge (CAS) facilities (e.g., an average of 10 to 12 hours compared to up to an average of 24 hours), it is prudent to apply a safety factor (10 to 20%) to the design equalization volume. The effect of flow equalization on MBR flows is illustrated in Figure 5.1

Design issues associated with flow equalization include confirming design storm wet-weather flows; sizing the equalization tank by establishing the allowable membrane flux and duration for treating equalized flows; and determining where flow equalization occurs (i.e., external tank, biological tankage, or both). An additional consideration is to examine the volume requirements associated with using the flow equalization basin to invert (to store temporary in the equalization basin) the dry weather diurnal flow to reduce energy costs. This cost can be significant in states like California, where peak demand energy costs can approach $1 per kilowatt-hour. Flow equalization tank-specific design issues to be considered include odor control, tank cover, and tank mixing and aeration (i.e., maintaining a minimum dissolved oxygen of 0.1 mg/L).

1.3.2 *Design Tradeoffs with Peak Flow Management Solutions*

Table 5.2 summarizes design tradeoffs associated with peak flow management facilities and MBR system sizing.

Return activated sludge (RAS) pump sizing is critical to membrane operation (Table 5.2). Indeed, if pumps are undersized, it could result in increased solids transport toward the membranes. In addition, this increased solids load on the membranes

TABLE 5.2 Peak flow management design decisions.

	Unit processes affected				
Peak flow management design issues to consider	**Membranes**	**Influent pumping station**	**External equalization tank**	**Equalization in biological tankage**	**Hybrid treatment system**
Size RAS pumps to provide adequate recycle at peak flow to maintain membrane filterability	X		X		X
Confirm HRT in pre-anoxic zones during peak flow + peak recycle conditions	X				
No grit removal facilities			X	X	X
Size to accommodate inverting dry weather diurnal to reduce energy cost		X	X		
Mixing/aeration			X		
Odor control			X		
Design for solids removal (e.g., tipping buckets for external equalization basin washdown)			X		
Biological aeration system				X	
Facility footprint	X		X	X	X
Peak flow system startup and shutdown			X		X

can result in intensive fouling of the membranes and operational upsets. Return activated sludge design criteria will vary if equalization or hybrid treatment is used as part of overall plant design. As shown in Table 5.2, the approach to peak flow management affects design decisions in different unit processes in addition to facility layout and, therefore, requires careful evaluation to ensure that unit process design functions as an integrated system.

1.4 Facility Hydraulic Gradeline and Hydraulic Design Issues

Establishing the hydraulic gradeline (HGL) and overall hydraulic design of the plant are two of the most important factors when designing wastewater treatment facilities. Key hydraulic design issues that need to be addressed include the amount of head required for the plant, especially if the MBR system will be designed for gravity permeating; treatment-train flow splitting (biological tanks and membrane tanks); and incorporation of large recycle flows (RAS and mixed liquor recycle [MLR]) back to the treatment train. Hydraulic design of biological tank baffling is detailed in Section 2.0 of this chapter. It is important to define the pumping configuration for the liquor (to or from the membranes) at the beginning of the project.

Before establishing the plant HGL, the underlying geology must be understood to determine if issues like rock, high groundwater, or differential settlement exist. Mitigation of any of these issues could become prohibitively expensive if not accounted for when establishing the plant HGL. Depending on the underlying geology, there may be cost savings associated with either an elevated structure for both the biological and membrane tankage or locating the biological and membrane tankage away from one another by pumping mixed liquor from the biological tanks to the membrane tanks. These design decisions can be accommodated by a MBR system design, given that the membranes have their own tankage separate from biological tankage. In addition, either one of these configurations may pump mixed liquor to the membrane tank (with gravity flow of RAS back to the biological process) or provide gravity flow from the biological tanks to the membrane tank (using pumps to transfer RAS back to the biological process). The design decision is a tradeoff involving structure costs, pumping costs (both capital and energy), piping costs, consideration for future expansions, and footprint restrictions. Another key design decision associated with HGL is whether the site can accommodate "gravity permeating," which uses the available head in the plant hydraulic gradeline to permeate water through the membranes without pumping. Depending on the fall across the plant site, gravity permeating may be a cost-effective option to consider. The designer needs to identify the geotechnical and layout constraints early in the overall design process.

Flow splitting to biological treatment trains and membrane treatment trains is a key design issue, given the large recycle flows (RAS and MLR) brought back to the treatment train. Depending on process configuration, RAS and MLR flows can total up to 6 times the average plant flow. Depending on the process and pumping configurations, the designer has two options for incorporating these flows back into the MBR treatment trains: dedicated pumps for each of the treatment train for each of the recycle flows or dedicated flow-splitting structures for each of the return flows, with weirs or flumes to ensure these flows are properly distributed to the treatment trains.

1.5 Pretreatment

This section addresses design and configuration of pretreatment components based on design decisions and criteria established during process design for the MBR system (see Chapter 4). Process-related design decisions include minimum fine-screen sizing, flow equalization size, whether or not grit removal is required as part of membrane equipment warranty, and whether or not primary clarification is part of the treatment process. In the context of these process decisions, a discussion of the facility design, configuration, and capacity of pretreatment systems is provided, together with a brief discussion of solids handling for these facilities. For a more detailed discussion of the design of solids handling facilities, refer to *Design of Municipal Wastewater Treatment Plants* (WEF et al., 2009).

1.5.1 Coarse Influent Screening

Membrane bioreactor membrane equipment suppliers require fine screening to protect their membranes because removal of solids from the influent wastewater is essential to MBR system operation. Because fine screening is required upstream of the membranes, two-stage screening is often used. With two-stage screening, upstream coarse influent screens protect the downstream fine screens from damage and blinding. This section addresses facility design issues associated with coarse influent screening as it affects MBR system design. For a more detailed discussion of coarse screening and available technology options, refer to *Design of Municipal Wastewater Treatment Plants* (WEF et al., 2009).

Membrane bioreactor suppliers specify the maximum opening size for fine screens. The typical 1- to 3-mm fine-screen size translates into a coarse-screen opening size ranging from 9 mm (.38 in.) to 3 mm (.13 in.). Coarse-influent screening design depends on the following conditions: the membrane equipment supplier-specified fine-screen opening size; the depth of the influent sewer (dictates coarse-screening technology options available); the need for screening upstream of the influent pump station (if required, allows for smaller coarse-screening size); use of equalization; and use of primary clarification between the coarse- and fine-screening unit process steps (also allows for a larger coarse-influent screen opening size). These issues need to be evaluated when selecting a coarse-influent screen technology and opening size. Other issues like odor control, indoor and outdoor screen location, screenings conveyance, and washing and compacting equipment require design consideration when laying out screening facilities. Coarse-influent screen-opening size tends to be smaller for MBR treatment facilities than for CAS system influent screening. Therefore, screenings volumes and processing rates for MBR coarse-influent screens will tend to be higher than for conventional treatment influent screens. For data regarding screenings volume as a function of screen opening size, refer to *Design of Municipal Wastewater Treatment Plants* (WEF et al., 2009). In addition, it is recommended that designers coordinate with screen equipment suppliers to confirm supplier experience with two-stage screening for similar applications. For a detailed discussion of screenings handling and the design of screenings handling facilities, refer to *Design of Municipal Wastewater Treatment Plants* (WEF et al., 2009).

1.5.2 Grit Removal

Grit removal system design considerations are similar for both MBR systems and CAS systems; therefore, this section will present only a brief discussion of the design issues. For a more detailed discussion of grit removal system design considerations, refer to *Design of Municipal Wastewater Treatment Plants* (WEF et al., 2009). Depending on whether or not grit removal is part of the MBR membrane equipment supplier's warranty, the grit removal function may be a conventional grit removal system (note that aerated grit systems should not be used if the downstream biological process begins with either an anoxic or anaerobic treatment step). If site-specific grit is not a significant concern during dry-weather flow, and the MBR membrane equipment supplier's warranty does not require an upstream grit removal requirement, the grit removal function may be provided as part of the flow equalization basin design or grit may just accumulate in the biological tankage. If grit removal facilities are required, determining the location of grit handling facilities (i.e., indoor or outdoor) and odor control need to be addressed.

1.5.3 *Fine Screening*

As stated previously, MBR membrane equipment suppliers require fine screening to protect their membranes. Typically, the fine-screen opening size is explicitly stated in the membrane warranty. In some instances, fine-screen equipment may be included in the MBR membrane equipment supplier's scope of services, given that some MBR membrane equipment suppliers require a specific fine-screen technology to ensure optimum membrane operation.

Common fine-screen equipment types used in MBR systems include center-feed band screens; rotary drum screens; rotary basket screens; and microscreens. Bar and wedge wire screens that may allow fibrous materials like hair to pass through are not as effective at protecting downstream membranes as other screen types, such as perforated-plate or mesh-type screens. Table 5.3 describes these fine-screen equipment types. Given the trend toward small fine-screen openings, designers should

Table 5.3 Fine-screening technology options typically used in MBR systems.

Fine screen	Description
Center-feed band screen	Uses rotating panels enclosed and submerged in a stainless steel tank and positioned parallel to the flow. The panels are attached to a roller chain that passes around an ultra-high molecular weight plastic guide to eliminate the need for submerged sprockets or bearing. Wastewater flows into the center of the screen and then passes through the screening panels on the left and right and through the bottom. Solids are collected on the panels and lifted to the discharge point where a spray system washes solids into the screenings washer/compactor or screw conveyor. Primary advantages of this type of screen are that wastewater has to flow through the screen panels and that there is no release of screenings downstream of the screen when the screen re-enters the water after screenings are lifted out of the wastewater channel. The primary disadvantages are that some of these band screens have reduced screenings removal by debris passing through gaps between panels and the gaps between the panels and the screen frame (the gaps on some screens are significantly larger than the screen perforations), and the failure of seals that some screens use to fill these gaps. Other disadvantages are that the flow must generally be pumped into the unit and that these screens use large amounts of water to flush the screenings out of the unit and transport them to a washer and compactor mechanism or screw conveyor, which must be sized to handle flushing water.

(*continued*)

Table 5.3 Fine-screening technology options typically used in MBR systems (*continued*).

Fine screen	Description
Rotary drum screens	Uses a slowly rotating drum with a screen basket. These screens also can be installed in a channel or in an enclosed steel tank (self-contained, typically for smaller flows). The screen basket can be made from perforated plate wire mesh or perforated plate screens (available in 2 mm). The water flows via gravity through the screen basket from the inside out. Floating and suspended materials larger than the perforations are retained, as more and more screenings are retained in the screening drum. When the upstream water level reaches a certain height, the drum begins to rotate. Cleaned perforations are rotated into the flow and retained screenings are rotated out of the flow. Screenings are flushed by spray water into a collection trough equipped with a screw conveyor to transport screenings out of the channel. These screens are typically equipped with an integral washer and compactor. The primary advantages of this type of screen are that it induces relatively low headloss, is mechanically simple, is designed to eliminate screening bypass, and can have lower headroom requirements because of its angled orientation. The primary disadvantages are brush maintenance because the screens rely on brushes to clean the perforations and that more floor area and wider channels may be required.
Rotary basket screens	Uses a stationary basket made of perforated plate. It uses a carbon steel shafted or shaftless spiral, which typically is fabricated in a continuous flight, to convey screenings out of the flow. The spiral rests within an enclosed steel trough or tube which minimizes odors. As the spiral turns, it clears screenings from the stationary basket and conveys them out of the channel. The bottom end of the spiral is fitted with cleaning brushes that also serve as the lower bearing. The advantage of this type of screen is its relatively low capital cost. The disadvantage is that the screening area is smaller, which could be problematic at high solids loadings and flow rates.
Microscreens	Uses a two-dimensional wire mesh with a clearly defined screen opening that removes fibrous material. Similar in design to a drum screen with a drum-shaped screen basket. The advantages of this type of screen include low headloss and lower headroom requirements because of its angled orientation.

coordinate with screen equipment suppliers to confirm supplier experience when sizing screenings handling systems. Required fine-screen capacity depends on whether or not upstream flow equalization is provided. Flow equalization can significantly decrease the cost of fine screens and screenings handling equipment; therefore, this should be taken into account when evaluating the cost-effectiveness of flow equalization for plant design.

1.5.4 *Package Headworks System*

Package headworks systems, consisting of coarse screening, grit removal, and fine screening, have been offered for MBR installations for the past few years. In situations where screening upstream of influent pumping is not required or the plant is significantly lower than the influent sewer, package headworks systems provide an alternative to conventional headworks. The primary benefits include reduced installation and startup time, better economics, and an approach that can be covered with odor control taps provided, allowing for outdoor installation, which eliminates the need for an odor-controlled building or enclosure.

Designers need to be aware that package headworks offerings are different enough to require the design approach to either accommodate the first two named suppliers or specify language in the package headworks informing the contractor that the design documents represent a system designed around the first-named vendor. If the contractor chooses a vendor other than the first named, any engineering required to modify the current design to accommodate the non-first-named vendor, and corresponding construction changes, shall be borne by the contractor.

1.5.5 *Primary Clarification*

The decision to include primary clarification in the treatment train will typically be made as part of the overall process design, as discussed in Chapter 4. Design considerations concerning the effect of primary clarification systems on the entire project are similar for both MBR systems and CAS systems; for more information, refer to *Design of Municipal Wastewater Treatment Plants* (WEF et al., 2009). As previously identified, key effects associated with adding primary clarification to the MBR treatment train are the reduction in 5-day biochemical oxygen demand to the biological process and the ability to locate the fine screens downstream of primary clarification. Primary clarification will reduce fine-screen screenings volume and may minimize the size of the fine screens when primary clarification is designed for hydraulic equalization of the water flow.

1.6 Design Considerations for New Membrane Bioreactor Facilities

This section addresses considerations unique to MBR facilities when selecting either a greenfield site or existing facility expansion with new tankage. These considerations, together with the process design (covered in Chapter 4), provide the basis for the MBR design discussion in Section 2.0 of this chapter.

All of the hydraulic-related considerations discussed in Section 1.4, particularly geotechnical considerations, need to be addressed when looking at a greenfield site. Other key issues include whether to enclose the facility in a building; how much room to provide for future expansions; and should the selected supplier's next generation of membrane requirements (tank-size requirements, lifting requirements, and increased capacity for the fixed footprint) be considered in the initial design. These issues invariably come down to both construction and life-cycle costs, which must be presented to the owner to obtain their approval of the design approach.

1.7 Design Considerations for Membrane Bioreactor Retrofitting Using Existing Bioreactors and/or Tanks

Design issues that need to be assessed when evaluating an opportunity for an MBR retrofit project include confirming that MBR hydraulics requirements can be met with the available grade head across the plant; confirming that membrane redundancy requirements can be met with existing tanks available for the retrofit, or that additional tanks can be constructed within the available footprint; confirming that membrane equipment and appurtenances can be accessed after the retrofit has been completed; and confirming that plan operations can be maintained during construction by sequencing of existing operation and identification of any temporary facilities. If retrofitting is to be performed on existing tanks, a structural evaluation of these facilities is required to determine what, if any, structural modification may be required to facilitate construction for membrane installation.

1.8 Design Considerations for Upgrading to the Next Generation of Membranes

Two different, yet related market factors that are influencing the MBR marketplace have direct ramification for MBR designers. These are (1) the expected time-to-market for the next generation of membranes, and (2) the potential for membrane replacement by suppliers other than the original supplier. Both factors have the following in common: for available membrane tankage, installation of new membranes in the existing tankage will increase membrane treatment capacity. Membrane bioreactor system designers and project managers need to be aware of these market drivers and, depending on project-specific circumstances, should consider what plant design decisions to recommend to the owner that can be implemented at minimal present cost to take advantage of the improved technology that will most likely be available in the near future.

If addressed initially, a number of design decisions would facilitate upgrade to the next generation of membranes. These decisions include provision of excess HGL; either oversizing or providing provisions to easily upgrade critical system components, including aeration equipment (or certain aeration-related components); biological or membrane tankage; pumping and piping facilities; membrane ancillary facilities; membrane hoist capacity; and certain electrical and control systems.

1.9 Other Design Considerations

Membrane bioreactor system design considerations for cold weather conditions (e.g., effects on sustainable flux rate and required membrane area and the potential for having to enclose membrane facilities) are site-specific and should be coordinated with the supplier based on their experience in these types of conditions.

2.0 MEMBRANE BIOREACTOR FACILITY DESIGN PRINCIPLES AND CONCEPTS

Issues related to the design of MBR bioreactor layout, equipment configuration, MBR aeration, and membrane permeation are reviewed in this section. The goal of this section is to provide relevant design criteria and recommendations regarding system redundancy and controls.

2.1 Membrane Bioreactor System Layout

Although the biological process for an MBR system is similar to that of a CAS system, there are some differences that affect bioreactor tank design and process pumping configurations. These differences are reviewed in this section.

2.1.1 Biological Tank Design Issues Specific to Membrane Bioreactors

Unlike the design for most CAS systems, biological process design for MBRs typically does not account for the need to generate a well-settling sludge or to minimize scum or foam. However, the membranes themselves do not allow foam or scum to pass through, effectively trapping them in biological and membrane tanks. Accordingly, it is important to make allowances to deal with foam accumulation in the tanks. Antifoaming agents can be used, such as a chlorine solution spray; however, it is not always desirable to rely on chemical addition. Providing sufficient freeboard to accommodate some foam accumulation is another strategy. Recent MBR designs have provided the ability for surface wasting, which has been shown to be effective in minimizing foam accumulation in both MBR and CAS systems. The general approach is to waste sludge from bioreactor tanks by allowing mixed liquor suspended solids (MLSS) to flow over a weir into a sump. Accumulated MLSS (or waste activated sludge) is then pumped out of the liquid treatment process to solids handling facilities. Because scum and foam accumulate at the surface of the tanks, this approach allows for continual removal of scum and foam as MLSS is wasted from the system.

As discussed in previous sections, most MBRs will be designed with anaerobic or anoxic zones at the front of the biological reactor tanks to provide for biological nutrient removal. Chapters 3 and 4 provide further information on the configuration of MBRs for biological nutrient removal and on process design strategies for minimizing the effects of the returning highly oxygenated RAS streams. The high rate of RAS or recycle streams inherent to MBR designs make MBRs more susceptible to short-circuiting. Strategies for addressing short-circuiting include providing sufficient mixing energy and configuring the zone, inlet, and outlet geometry to ensure good flow distribution. Use of hydraulic modeling for the design of bioreactors in MBR processes is becoming more commonplace, and the use of hydraulic models, including computational fluid dynamic models, should be considered where appropriate to evaluate and optimize tank geometry. Another consideration in the design of biological reactors is provision of sufficient freeboard. Most bioreactor tanks in MBR systems are designed for 91 cm (3 ft) or more of freeboard. This allows for some room to accommodate foam, but also provides for the flexibility to operate over a wide range of hydraulic conditions that are encountered because of the wide range of RAS rates typically used. Some designs also use bioreactor tanks for flow equalization, which could require additional freeboard depending on the amount of storage volume needed. Using bioreactor tanks for flow equalization means the water level in the reactors will be variable; as such, the freeboard, pumping, and aeration systems need to be designed accordingly. Aside from providing for sufficient freeboard, use of in-tank equalization has a significant effect on aeration blowers and requires that they are of a type that can accommodate a variable-discharge pressure.

The bioreactor tanks should also be designed so that they can be taken out of service if needed for maintenance or grit removal. Like CAS bioreactor tanks, most MBR tanks are designed with sloping floors and tank drains to facilitate emptying tanks for maintenance.

2.1.2 *Membrane Tank Shape and Redundancy*

Membrane tank shape is largely driven by the size and geometry of the membranes. Given the hydraulic limitations inherent in an MBR system, special attention is needed to ensure adequate redundancy in the system for chemical cleaning (and unplanned maintenance). While some facilities are designed for a full standby train, others are not, depending on the nature of permit limits and the owners' risk management policies. Facilities without standby trains will typically have contingency plans to store or divert flow from the membrane system should any equipment be out of service during peak wet-weather periods. It is also important to make provisions for extra space for membrane cassettes to accommodate future growth or the possibility that the system's actual sustainable flux is less than design. The amount of extra space to be provided is project-specific; however, a survey of several consulting firms indicates common practice is to provide anywhere from a 10 to 50% allowance for these considerations.

2.1.3 *Pump and Return Activated Sludge Configuration*

Return activated sludge in MBR processes provides the unique function of controlling the MLSS concentration in the membrane tanks, thereby helping to maintain membrane flux, particularly during peak flow events. During periods of peak flow, the rate of solids transferred from the bioreactor tanks to the membrane tanks increases, which results in an increase in the MLSS concentration in the membrane tanks. The increase in MLSS concentration in the membrane tanks can be reduced somewhat by increasing the RAS; however, there are practical limitations to how much the RAS rate can be increased. The magnitude of the MLSS increase will also depend on the size of the membrane tank as this effect is more pronounced as the membrane tank size is reduced.

Accumulation of excess solids in the membranes is quickly exacerbated when there are imbalances in flow splitting or RAS flow distribution. This highlights the importance of achieving good flow distribution, which can be achieved by having dedicated pumps for each tank, providing a hydraulic splitter box and structure, or having another way to positively control split to different tanks.

The following sections summarize some of the key elements of the two pumping and flow-routing configurations for RAS. As described in the previous sections, the two approaches are (1) to use gravity flow to transfer mixed liquor from the bioreactors to the membrane tanks and pump RAS back to the bioreactors, and (2) to pump mixed liquor from the bioreactors to the membrane tanks with gravity flow for RAS. While there are subtle differences between the two approaches, the main driver behind selecting an approach is typically the one that best fits the constraints of the plant site and HGL. The difference in energy use between the two approaches is proportional to the percent difference between the MLSS concentrations in the bioreactors and the membrane tanks; however, the absolute magnitude of the difference in annual power costs between the two pumping configurations is relatively small. For example, the difference in power costs between the two configurations for a 10 mgd facility, with MLSS concentrations in the aeration and membrane tanks of 8000 mg/L and 10 000 mg/L, respectively, is only about 20%.

2.1.3.1 Gravity Flow to the Membrane Tank and Return Activated Sludge Pumping Gravity flow of MLSS from the bioreactor tank to the membrane tank and pumping RAS from the membrane tanks back to the bioreactor tank is one common configuration. In this configuration, the pumping system is sized for the recirculation ratio (R) times the influent flow (Q). With this configuration, the RAS flowrate is typically controlled by trying to maintain a flow setpoint as a percentage of Q. The main advantage of this approach is that it is simple and minimizes the amount of process flow pumping between the bioreactor and membrane tanks.

2.1.3.2 Pumping to Membrane Tank and Return Activated Sludge Gravity-Flow Return Tank The other approach for returning RAS to bioreactor tanks is to use gravity flow. In this configuration, flow from the bioreactors must be pumped to the membrane tanks; therefore, the pumping system must be sized for (R + 1) times the influent flow. With this configuration, the pumping flowrate can be controlled by maintaining a flow setpoint or, alternately, with level control in the bioreactor tanks. In either instance, RAS flow from the membrane tanks will typically flow over a weir by gravity and return to the bioreactor tanks. This configuration requires more pumping between the bioreactor and membrane tanks than the previous approach; however, there is a benefit. By pumping up to the membrane tanks, it is typically more feasible to implement gravity permeation as opposed to being required to pump permeate. Therefore, this approach may offer the possibility of eliminating or minimizing use of the permeate pump station.

2.2 Membrane Bioreactor Aeration

Process aeration and membrane air scour are critical components of the bioreactor and membrane tanks, respectively. Design considerations for these facilities are reviewed in this section.

Aeration for the bioreactor is typically provided by the same type of blowers and diffused aeration systems used in CAS systems. The primary difference is that MBR systems typically are more heavily loaded and, consequently, have higher oxygen uptake rates than CAS systems. For this reason, MBR aeration systems typically consist of full floor coverage of high-efficiency fine-bubble diffusers, with membrane discs and tubes most commonly used. However, one issue to consider is having sufficient floor space for the required number of diffusers with deep tanks, high MLSS, high biochemical oxygen demand (BOD) loading, and long solids retention times (SRTs). Table 5.4 shows typical design parameters for aeration systems used in MBRs.

Figure 5.2 illustrates typical diffuser density in terms of area tank and area diffuser (AT/AD) (i.e., total floor area divided by diffuser area) as a function of different SRTs and peak-month BOD loading on the aerobic zone of a 4.6-m (15-ft) -deep bioreactor tank. As shown in Figure 5.2, increased BOD loadings and longer SRTs require more diffusers to meet process oxygen demands. Practical limits for most diffuser suppliers typically limit AT/AD values to a minimum of 2.5 to 3. Higher diffuser densities are difficult to achieve while maintaining sufficient area to allow periodic maintenance or replacement of the equipment.

TABLE 5.4 Typical design parameters for aeration systems used in MBRs.

Item	Description
Commonly used diffuser types	Membrane disc or tube
Typical design oxygen uptake rate, mg/L/hr	50 to 150
Diffuser AT/AD	3 to 6 typical (variable, see Figure 5.2)
Design operating dissolved oxygen (aerobic zones), mg/L	1 to 3

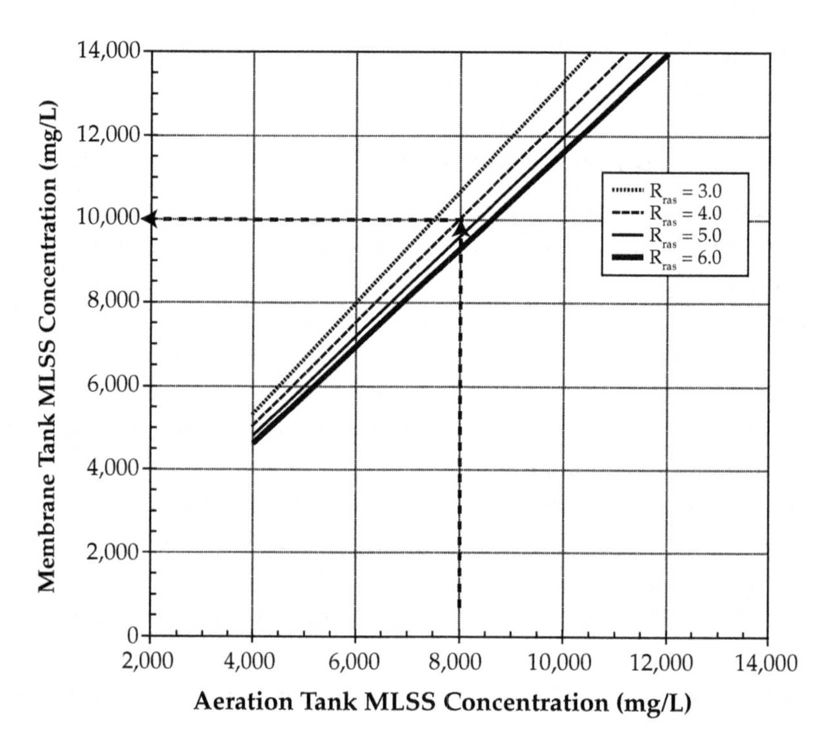

FIGURE 5.2 Typical diffuser density in terms of AT/AD as a function of different SRTs and peak-month BOD loading on the aerobic zone of a 4.6-m (15-ft) -deep MBR tank [lb/kcf (kcf = 1000 cu ft) × 16.02 = g/m^3].

2.3 Membrane Permeation

Permeate from the membrane tank can be removed by gravity, pumping, or a combination of both of these processes.

2.3.1 *Pumped Permeation*

For immersed membranes, pumping is the most common approach to generating the pressure differential needed across the membranes to draw water through the membranes. This is achieved by using pumps to apply suction to the clean water side of the membranes. Permeate pumping can be used for any type of membrane system. Based on typical flux rates, this requires a differential pressure of 0.3 to 7.0 psig across the membrane, including losses in piping and fittings. Pumping systems must be designed to accommodate the low net positive suction head conditions on the pump suction side, providing the required negative pressure in the range 0.3 to 7.0 psig.

2.3.2 *Gravity Permeation*

For immersed-style membranes, gravity permeation can be achieved if there is sufficient drop in the HGL between the membrane tank and the next downstream process tank. A differential pressure similar to what is needed for pumped permeation systems is also needed for gravity permeation. This means the HGL must be able to accommodate up to 6.1 m (20 ft) of loss.

2.3.3 *Pump-Enhanced Gravity Permeation*

Pump-enhanced gravity permeation consists of using pumps when there is not enough head when membranes are dirty or when additional capacity is needed. In this situation, the MBR is operated in gravity mode when membranes are clean, whereas permeate pumps start when there is no sufficient gravity pressure to push water through the membranes as they start fouling.

3.0 ANCILLARY FACILITIES AND EQUIPMENT DESIGN CONCEPTS

3.1 Membrane-Related Ancillary Facilities

3.1.1 *Clean-in-Place System*

The clean-in-place (CIP) system is specific to the membrane supplier. Depending on the membrane system type, the CIP system may be more or less complex. The CIP system for hollow-fiber membranes typically includes storage tanks and metering pumps for CIP chemicals, a CIP tank for preparation of the CIP solution, and a pump to deliver and/or recirculate the solution. Clean-in-place chemicals for hollow-fiber membranes include sodium hypochlorite for removal of organic fouling and an acid for removal of scaling.

Diaphragm-type metering pumps are used for delivering the chemicals to the CIP tank. Cross-linked high-density polyethylene tanks are typically used for the CIP tank. However, fiber-reinforced plastic (FRP) tanks are sometimes used for smaller systems. Although FRP tanks are more durable and hold up better to high chemical concentrations, they are also more expensive. Air agitation is sometimes provided in the CIP tank for mixing.

The CIP system for flat-plate membranes is less complex. The system typically includes bleach stock, a pressurized water line, and an eductor. The bleach stock is often stored in a drum. The solution is entrained into the pressurized water through the eductor and conveyed to the membranes for chemical cleaning.

3.1.2 *Permeate and Backpulse Pumps*

Centrifugal pumps are commonly used for permeate pumps in MBR systems. The pump is situated near the base of the tank, creating a local high point in suction piping; in addition, an air–water separator may be required between the membranes and the permeate pumps to keep vacuum pressure and protect the pump against cavitation, depending on the vertical profile of the system and pump locations relative to membrane locations. Self-priming pumps have also been used for permeate pumps. When using self-priming pumps, the pumps are situated above the water surface elevation in the membrane tank and no air–water separator is required. Self-priming pumps are, however, more expensive and more complex to maintain compared to centrifugal pumps.

The first generation of MBR systems used the permeate pump for backpulsing. The permeate pump, which was typically a centrifugal pump, can pump flow in a single direction only. The reversal of permeate flow for a membrane backpulse was initially accomplished by configuring additional piping and valves around the permeate pump to permit direction of permeate back through the membranes, as shown in Figure 5.3(a). The additional piping and valves require a larger footprint and can be quite costly for medium-to-large-sized treatment plants. To reduce footprint and cost, the next generation of MBR systems used a separate pump for backpulsing. Each membrane train shared a common backpulse pump system, as depicted in Figure 5.3(b). While the separate backpulse pumps reduced the number of valves, it introduced another set of pumps requiring space and maintenance. Rotary lobe pumps, which can pump flow in both directions, provide an alternative to centrifugal pumps with complex piping and separate backpulse pump systems for accomplishing permeate pumping and backpulsing. Figure 5.3(c) shows the piping configuration with rotary lobe pumps. Rotary lobe pumps are, however, more expensive and have a lower flow capacity than centrifugal pumps. Owner preference, cost, space constraints, and treatment train capacity are key considerations in determining which type of pumps to use.

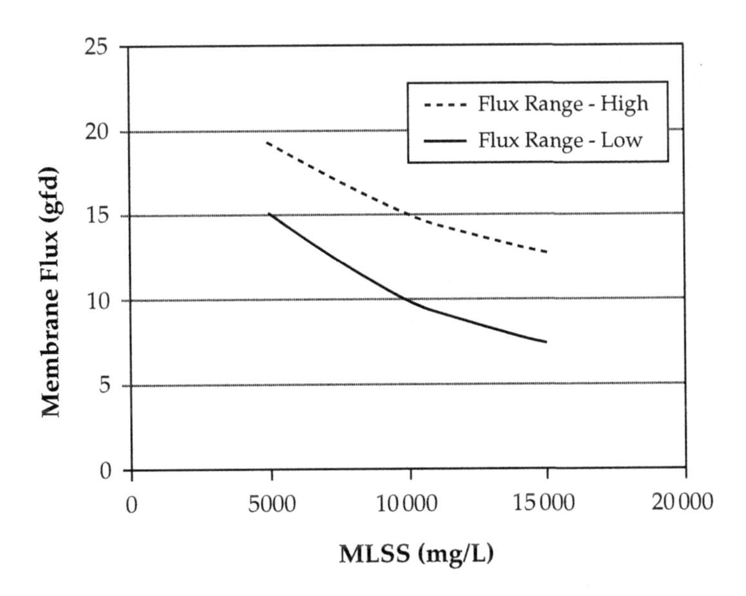

Figure 5.3 Backpulse pump configurations: (a) valves and piping configured to permit using permeate pump for backpulse, (b) separate backpulse pumps, and (c) rotary lobe pumps that permit flow in both directions.

3.1.3 *Service Air*

Most MBR systems are designed with pneumatically actuated valves, which are more corrosion-resistant, more reliable, less expensive, and respond faster than motor-actuated valves. Pneumatically actuated valves require service air. The equipment for service air includes air compressors, air receivers, oil–water separators, air filters, and air dryers (in cold weather).

3.1.4 *Air Scour Blowers*

Membrane system blowers supply air for scouring of the membrane surface. Centrifugal blowers, which have high efficiencies, are commonly used but are not well suited for applications with variable liquid levels. Positive-displacement blowers may be used if fluctuations in the water surface level in the membrane tank are anticipated, but they are much less efficient. The membrane system blower is typically within the scope of equipment provided by the membrane supplier and is sized by the membrane supplier.

3.2 Bioreactor Aeration Blowers

Bioreactor aeration blower system sizing is different for MBR systems, relative to CAS systems, based on the high MLSS and corresponding low alpha factor for MBR systems. Energy consumption for bioreactor aeration contributes to a significant portion of the energy requirements at a MBR wastewater treatment plant (WWTP). Therefore, one of the primary considerations in bioreactor aeration blower selection is energy efficiency. Centrifugal blowers and positive-displacement blowers are commonly used for bioreactor aeration. Centrifugal blowers are typically used for larger systems because of their high efficiency, but they come at a relatively high capital cost. Positive-displacement blowers are less expensive, but they are also less efficient. Turbo blowers are reportedly even more efficient than centrifugal blowers, but they are a newer technology and more costly.

If the bioreactor tank is used for equalization, then the aeration system must be designed to account for fluctuating water-surface levels. Positive-displacement blowers maintain relatively constant capacity with changing pressure and are, therefore, more suitable for operating with fluctuating tank levels. Single-stage centrifugal blowers with dual control of flowrate and pressure can also be used in applications with tank level fluctuations. One manufacturer of single-stage centrifugal blowers has both an inlet and diffuser guide vane that permits control of both the flowrate and pressure developed by the blower. Multistage centrifugal blowers can be used by using an inlet throttling valve or variable frequency drives. Capital and energy costs are two key considerations in the selection of a blower type.

When designing the bioreactor aeration blower, noise and sound reduction should be considered. Blowers should be supplied with decibel-control enclosures, particularly if they are located outside. Other considerations include specification of proper coating for corrosion control if the blower is also being used to pull foul air.

3.3 Pumping Facilities (Return Activated Sludge, Mixed Liquor Recycle, Waste Activated Sludge, Mixed Liquor Transfer, and Scum)

The type of pump used for these applications depends on the size of the facility. Table 5.5 lists several different types of pumps and their typical service uses. For smaller facilities, immersed or dry pit pumps can be used. Dry pit pumps are more commonly used at medium-to-large facilities as they are more accessible for servicing.

Table 5.5 Different types of pumps and their service uses.

Pump type	Service	Application
Submersible pump	RAS, MLR, WAS, MLT	
Propeller pump	RAS, MLR, WAS, MLT	For high-flow, low-head applications
Horizontal close-coupled centrifugal pump	RAS, MLR, WAS, MLT	
Vertical close-coupled centrifugal pump	RAS, MLR, WAS, MLT	Used when space is a constraint, but less reliable than other pump types because of how the motor is mounted
Positive displacement	Scum	

3.4 Anoxic and Anaerobic Basin Mixers

When MBR systems need to provide deep nitrification and denitriciation, they are designed with anoxic and/or anaerobic basins. Because MBR systems have higher RAS recycle rates and smaller bioreactor tanks than CAS systems (and, therefore, shorter HRTs), mixing in the anoxic and aerobic zones is critical to treatment performance. The mixer must be able to mix at the bottom of the tank. Submerged mixers on rails are commonly used because they are relatively inexpensive and effective. Shaft mixers on platforms are also sometimes used.

3.5 Cranes and Hoists

Cranes and overhead hoists, or other types of mobile lifting equipment, are needed to remove membranes for servicing, repairs, or replacement. The equipment should be sized to handle the weight of wet and "sludged" membranes (i.e., a membrane with wet sludge stuck to its surface) and incorporate additional handling capacity for the potential weight of future generations of membranes. Future generations of membranes will likely be denser as membrane manufacturers seek to increase filtration area within a smaller space. The sizing of membrane lifting equipment should include an additional 20 to 25% lifting capacity for future generations of membranes.

3.6 Membrane Bioreactor Tank-Related Considerations

3.6.1 Dip Tanks

Dip tanks were designed for membrane cleaning operations for the first generations of MBR use. Performing membrane cleaning operations in a separate tank was cumbersome, and MBR membrane systems are now designed with a CIP system. Generally, dip tanks are not designed anymore. However, some operators prefer to have a dip tank at their facility for membrane repair and service. Other operators prefer to have a washdown pad to allow membrane cleaning for inspection, maintenance, or upgrade purposes, and to capture solids and debris to return to the plant or for disposal. During the design process, the owner and operations staff should be made aware of the cost and benefits of a dip tank and/or washdown pad for facilitating maintenance.

3.6.2 *Tank Design and Dimensioning*

In MBR systems, the membrane tanks are a relatively small volume compared to biological tanks and their volume varies among different MBR brands. However, membrane tanks experience an enormous hydraulic load from the large MLSS flowrate, which is in the range of 4 to greater than 6 times the filtrate flowrate. High flow through the relatively small volume of the membrane tanks may create significant challenges if the tanks are not designed and operated properly. As a result, a commensurate amount of attention needs to be given to designing membrane tanks. The absence of industry standards for low-pressure membranes (i.e., microfiltration and ultrafiltration), and the dual direction taken by MBR membrane suppliers in promoting both hollow-fiber and flat-plate membranes, complicates this issue even further.

For MBR systems with dedicated tanks for membranes, dimensioning of the tanks should be designed in cooperation with the membrane supplier who will be providing membranes for the particular project to ensure that tank geometry does not create hydraulic flow patterns or areas of deposition that could interfere with air scouring of the membranes or allow hydraulic short-circuiting. When different membrane suppliers are introduced to a project, the design needs to accommodate the specific requirements of the membranes offered by each supplier. Air scour systems pump large flows of water from the bottom to the top of the membrane tank through the membranes; as such, sufficient space must be provided around the membrane units, racks, or cassettes to allow this flow to return unimpeded to the bottom of the tanks while keeping uniform flow patterns around all the membranes in each tank. Because almost all membrane suppliers have module geometry and dimensions that are unique, generalization and a standard design approach are hard to achieve; however, certain common features of membrane tanks must be considered by the designer. These can be grouped into features associated with redundancy, tank hydraulics, piping, and access to membranes, valves, and fittings.

In conjunction with the selection of design membrane area, the designer must initially establish the number of units, racks, or cassettes per membrane tank and the number of membrane tanks deemed necessary to provide the desired level of reliability and flexibility of operation for the entire system. Reliability is, in large part, a policy decision of the plant owner that is influenced by the type of facility, the ability to bypass flow around the membranes, seasonal variability in permit conditions, and the occurrence and duration of peak flows and loads and minimum water temperatures. For example, a satellite treatment plant may be able to reduce the influent flow received by the plant or to cease operation entirely for repairs, and thus requires minimal redundancy. Similarly, a facility that is allowed to bypass peak wet-weather flows with physical and chemical treatment of the bypass (e.g., enhanced high-rate clarification) may also be able to treat design flows with minimal redundancy. Conversely, an end-of-the-line facility, without the ability to bypass, will require full $N + 1$ or $N + 2$ redundancy with N being the number of tanks or cells needed in service to treat design peak (maximum) flows. Specifically, this means that the plant needs to treat the maximum flow when one or two membrane trains or cells are out of service. The number of units, racks, or cassettes per tank will be established by module size and the practical aspects of sizing the scour air and permeate piping, valves, pumps, and blowers. Consideration should be given to the economics of designing fewer tanks with larger-capacity units or cassettes compared to the larger number of tanks required with smaller-capacity membrane units. As an example, the practical maximum for hollow-fiber membrane design is about 12 membrane units (cassettes) per tank. The length-to-width ratio of the membrane tanks can be altered to a limited extent by choosing to arrange the racks or modules in one or two rows. Having more than two rows unduly complicates tank piping.

Often, designers will leave space in the membrane tanks for additional units, racks, or cassettes to be added in the future as is necessary or desired. Alternately, one or more entire membrane tanks can be constructed, but left unequipped. Most commonly, designs provide free space to allow expansion of the membrane area by about 25%. This is typically done for one of two reasons. The primary reason is to facilitate and expedite future expansions in treatment capacity to meet growth in the service area without having to pour concrete. The second reason is to provide a margin of safety should the plant not be able to perform at the design flux or should hydraulic peaks be greater than design. The empty tank can be used for membrane cleaning, sludge collection, or other purposes until expansion is required.

Removal of permeate along the length of a membrane tank results in an increase in MLSS concentration so that a concentration gradient exists from the feed to the effluent points. To a large extent, effects of the concentration increase are minimized by the high-return recycle ratio and the intense energy input from the air scour system. Some designs minimize the concentration gradient in MLSS by providing longitudinal feed and effluent channels along the long dimension of the tanks or by feeding the tank in two locations (i.e., at each end with a center withdrawal or to the center with one-half of the feed directed to an overflow weir at each end of the tank). As mentioned previously, the hydraulic elements must be designed to provide a relatively even flow split to the membrane tanks in service, and the hydraulic design must account for the eventuality of one or more membrane tanks being out of service. Some membrane designs are vulnerable to physical damage when feed is first directed to an empty tank so tank inlets must be configured to avoid physical damage to the membranes from hydraulic surges that occur when filling tank. A simple approach is to design a baffle wall, barrier, or reflector at the feed pipe. Designs must eliminate the potential for scum or froth to be trapped in the membrane tanks. Air and permeate piping may be placed within the membrane tanks to minimize obstructions above the tanks, or above the tanks to facilitate access to piping and valves for maintenance. With either approach, the tanks must accommodate the necessary piping and valves and provide a reasonable degree of freeboard to minimize splashing and spray above the tanks. All fittings and accessories, such as valves, that need to be accessed during operation need to have easy access and must be located above the water level or in dry areas. Similarly, membrane tanks must be accessed at intervals for servicing membranes and tank design should facilitate expected maintenance requirements to the extent practical, including providing the ability to remove membranes without interference from piping and valves.

3.6.3 *Tank Grating and Covers*

Bioreactor tanks are typically not covered at mid-to-large-size facilities because of the size of the tanks. However, coverage of the membrane tank should be considered. Grating is typically provided over membrane tanks to release the large amount of air added to membrane tanks, to protect the membranes from objects falling into the tanks, to ensure worker safety, and to facilitate access above the tanks. Movable covers in lieu of grating have been used in some designs. Movable covers have the advantage of minimizing noise and eliminating spray. Membrane tanks can be left open to the atmosphere, covered by canopies, or enclosed in buildings, depending on the climate and aesthetic requirements of the site. Enclosed designs should take into consideration the release of potentially harmful gases during chemical cleaning of the membranes, the greater potential for corrosion from both excessive humidity and the presence of corrosive gases, and the ability to provide convenient and easy removal of membranes for routine maintenance by operational staff. Tank covers need to be equipped with proper ventilation as significant temperature increases through use of diffused-air systems can be observed, depending on the type of covers.

3.6.4 *Tank Coatings*

Typical design practice for municipal wastewater plants involves coating concrete tanks to prevent corrosion and increase tank life. In addition to the potential for corrosion based on the corrosive nature of wastewater, MBR membrane tanks can have the additional requirement for inplace membrane cleaning, which can involve addition of acidic or caustic chemicals to the membrane tank. Depending on the type of cleaning performed and membrane supplier cleaning requirements, the range of chemical dosage can vary several orders of magnitude. System designers need to be aware of the environments that will be experienced in MBR tanks and work closely with coatings vendors to identify cost-effective options that will increase tank life. It is important to bring in an engineering corrosion expert early on in the design process to examine water quality issues and understand the chemical cleaning regimen (i.e., anticipated frequency and duration under which dosages) to properly specify a coatings system for the MBR membrane tank.

A traditional epoxy coating may not be adequate to protect against corrosion. A polyurethane lining is better suited for the corrosive environment in an MBR membrane tank. Polyurethane liners are typically sprayed on in at least three coats and can be as much as 13-mm (0.5-in.) thick. T-Lock polyvinyl chloride (PVC) liners, which are commonly used to rehabilitate manholes and repair grit chambers, offer even more robust protection. However, T-Lock PVC liners are expensive and may be cost-prohibitive. Regardless of the lining used, careful application and inspection of the lining is critical.

In addition to corrosion protection, if the plant is in an area where seismic activity can occur, then a coordinated design effort to examine seismic requirements and how these requirements will work with coatings requirements needs to be evaluated to specify a system that will both provide corrosion protection and hold up during seismic events. Polyurethane linings can flex and tend to be suitable for seismic conditions.

3.7 Instrumentation and Control

A membrane supplier is liable for meeting performance specifications. To ensure that correct protocols, which are vendor-specific, are followed, the membrane supplier typically provides all related instruments, the programmable logic controller (PLC), and other controls for the systems. In addition to control requirements for the membrane system, other processes such as influent, sludge, chemical systems, and effluent also require control systems that need to be integrated with the membrane system. The membrane supplier's control system can be integrated with the plant's overall computer monitoring system, often referred to as the *supervisory control and data acquisition* (SCADA) system. Alternatively, the membrane supplier's PLC and control system can be set up as the master control system, and other processes requiring control can be incorporated into the membrane supplier's control system. Figure 5.4 shows an example of a SCADA system architecture where the membrane control system is a part of an overall control system.

Existing control systems, owner preference, and plant size and complexity will dictate the control scheme and control architecture. Upfront coordination with the

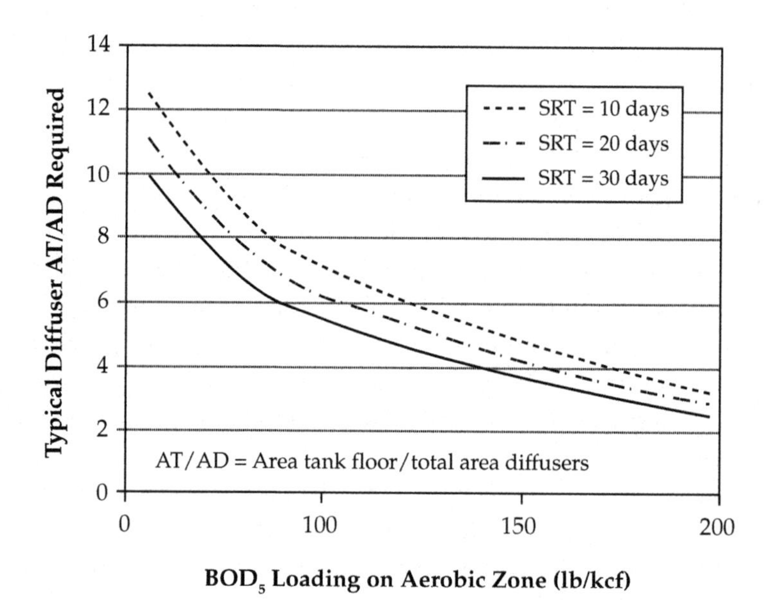

FIGURE 5.4 Diagram of sample SCADA system architecture showing overall components.

membrane supplier on the control system is important. Features of the SCADA system should include use of proven technology, user friendliness, expandability, and remote access.

4.0 SURVEY OF CURRENT MEMBRANE BIOREACTOR PRACTICES IN SIGNIFICANT DESIGN FIRMS

As part of the effort to collect information about current design practice for MBRs, a survey was developed and distributed to several consulting engineering firms with experience designing and operating MBR systems. Responses were obtained from four firms, which represent a significant percentage of the MBR facilities designed in the United States. Where appropriate, results of the survey were incorporated into various sections of this manual. A summary of the results are provided in this section.

4.1 Survey Size

Table 5.6 summarizes information about the number, size, and status of MBR projects that the responding consultants have designed. The responders have designed a total of 56 MBR facilities, with half of them currently in operation. The majority of facilities are less than 20 000 m^3/d (5 mgd); however, almost 40% of the facilities are greater than 20 000 m^3/d (5 mgd), and 15% of the facilities are larger than 75 000 m^3/d (20 mgd).

TABLE 5.6 Summary information about the number, size, and status of MBR design projects.

	Responding consulting/design firm				
Item	**A**	**B**	**C**	**D**	**Total**
No. of MBR facilities designed	**6**	**30**	**12**	**8**	**56**
Small facilities < 20 000 m^3/d (5 mgd)	4	15	10	5	34
Medium facilities—20 000 m^3/d (5 mgd)–75 000 m^3/d (20 mgd)	2	8	1	2	13
Large facilities > 75 000 m^3/d (20 mgd)	0	7	1	1	9
No. in operation	**2**	**15**	**7**	**4**	**28**

4.2 Reasons for Selecting a Membrane Bioreactor Process

Although the cost of implementing an MBR was typically compared to that of other technologies, the overwhelming response was that MBRs were selected for implementation primarily because of their compact footprint and high level of effluent water quality. Low cost was never identified as the primary driver for selecting an MBR at the time data were collected for this manual.

4.3 Project Delivery and Membrane Procurement

For the majority of projects (almost 75%), membranes were procured by the owner separately from any construction contract with a general contractor. Approximately 25% of the projects were implemented through a design-build delivery method. Only 1 out of 44 projects was implemented with a design-build-operate strategy.

4.4 General Configuration

Most responders did not specify a preference for the type of membrane (hollow fiber or flat plate), although one of the responders preferred hollow fiber. Most responders typically cover the membrane tanks, sometimes for odor control or to protect membrane fibers and equipment. Flow equalization is typically provided somewhere in the flow stream, and one-half of the responders typically tried to incorporate some flow equalization in the bioreactor tanks themselves. Primary clarifiers were sometimes incorporated in the flow trains at larger facilities, although they were rarely used for smaller facilities.

4.5 Pretreatment

All of the responders incorporated fine screening as a necessary pretreatment for an MBR system, although requirements varied from 1 to 3 mm. The need to implement coarse screening upstream of the fine screens varied depending on vendor and project constraints. Some responders recommended the use of 6-mm coarse screening upstream of any fine-screening facilities.

4.6 Membrane Flux

Typical design fluxes cited by responders ranged from 17 to 25 L/(m^2·h) (10 to 15 gfd) under average conditions, and 25 to 50 L/(m^2·h) (15 to 30 gfd) under peak-hour flow conditions for hollow-fiber membranes. Some responders noted that while they have internal guidelines, some flexibility is allowed to suppliers to allow for optimization of their system design.

4.7 Mixed Liquor Suspended Solids and Solids Retention Time

For the bioreactor tank, the maximum allowable MLSS ranged from 6000 to 10 000 mg/L and, for the membrane tank, the maximum allowable MLSS ranged from 10 000 to 12 000 mg/L. Many firms responded that the design SRT range was typically based on treatment objectives; however, some firms indicated a minimum SRT requirement of 8 to 10 days.

4.8 Return Activated Sludge Pumping

Typical RAS pumping flow requirements ranged from 2 to 6 times the influent flow (Q) under average or maximum monthly flow conditions.

4.9 Freeboard

Typical freeboard requirements for bioreactor and membrane tanks ranged from 0.6 to 0.9 m (2 to 3 ft). The minimum and maximum freeboards identified were 0.3 and 1.8 c (1 and 6 ft), respectively.

4.10 Redundancy

Responses varied widely on the level of redundancy required for an MBR system. Responders recommended having anywhere from no redundancy, the ability to take one train offline during dry-weather conditions, and simply having two bioreactor and MBR trains so that one could be taken out of service. Responders also indicated that there should be room in the tanks to add anywhere from 10 to 50% of the membranes and cassettes. Responders also indicated that RAS pumping should have either an installed standby pump or a shelf-spare available. As expected, redundancy requirements will generally be dictated by the cost and level of risk the owners are willing to accept.

4.11 Construction Materials

Most responders indicated stainless steel was the preferred construction material for most of the piping and submerged equipment, although PVC or ductile iron pipe has also been used. Chlorinated polyvinyl chloride or PVC were primarily used for the cleaning solution and the chemical system.

5.0 REFERENCES

Ongerth, J. E. (1979) *Evaluation of Flow Equalization in Municipal Wastewater Treatment*; EPA-600/2-79-096; U.S. Environmental Protection Agency, Municipal Environmental Research Laboratory, Office of Research and Development: Cincinnati, Ohio.

Water Environment Federation; American Society of Civil Engineers; Environmental and Water Resources Institute (2009) *Design of Municipal Wastewater Treatment Plants,* 5th ed.; WEF Manual of Practice No. 8; ASCE Manual and Report on Engineering Practice No. 76; McGraw-Hill: New York.

6.0 SUGGESTED READINGS

Alexander, K.; McBride, B.; Jackson, R.; Wade, J. (2001) Membrane Bioreactor Design: Problems and Solutions for a Plant Upgrade in Anthem, Arizona. *Proceedings of the 74th Annual Water Environment Federation Technical Exposition and Conference* [CD-ROM]; Atlanta, Georgia, Oct 13–17; Water Environment Federation: Alexandria, Virginia.

Benedek, A.; Côté, P. (2003) Long Term Experience with Hollow Fibre Membrane Bioreactors. *Proceedings of the International Desalination Association Conference;* Bahamas; International Desalination Association: Topsfield, Massachusetts.

Boe, R.; Quan, C.; Whitaker, B.; Klein, A.; Goodwin, S.; Crawford, G. (2006) Redundancy and Reliability—the 5th Generation of Membrane Bioreactor Design. *Proceedings of the 79th Annual Water Environment Federation Technical Exhibition and Conference* [CD-ROM]; Dallas, Texas, Oct 22–25; Water Environment Federation: Alexandria, Virginia; pp 1917–1918.

Brindle, K.; Stephenson, T. (1996) The Application of Membrane Biological Reactors for the Treatment of Wastewaters. *Biotechnol. Bioeng.,* **49** (6), 601.

Churchouse, S.; Brindle, K. (2003) Long Term Operating Experience of Membrane Bioreactors. *Proceedings of the 4th International Meeting on Membrane Bioreactors for Wastewater Treatment;* April 9; School of Water Sciences/Water Biotreatment Club, Cranfield University: Beford, U.K.

Churchouse, S.; Warren, S.; Floyd, M. (2007) Feedback from the Porlock MBR Plant in Its 10th Year of Operation: An Analysis of the Flux, Effluent

Quality and Membrane Lifetime Data to Date. In *7. Aachener Tagung Wasser Und Membranen;* Aachen, Germany, Oct 29–31; The Institute for Sanitary Engineering and Technology at the RWTH Aachen and the Society for the Promotion of Environmental Engineering: Aachen, Germany.

Cornel, P.; Krause, S. (2006) Membrane Bioreactors in Industrial Wastewater Treatment—European Experiences, Examples, and Trends. *Water Sci. Technol.*, **53** (3), 37.

Crawford, G.; Thompson, D.; Lozier, J.; Daigger, G.; Fleischer, E. (2000) Membrane Bioreactors—A Designers Perspective. *Proceedings of the 73rd Annual Water Environment Federation Technical Exhibition and Conference* [CD-ROM]; Anaheim, California, Oct 14–18; Water Environment Federation: Alexandria, Virginia; pp 311–319.

DeCarolis, J.; Oppenheimer, J.; Hirani, Z.; Rittman, B. (2009) Investigation of MBR Effluent Water Quality and Technology—A Worldwide Survey. *Proceedings of the 24th Annual WateReuse Symposium*; Seattle, Washington, Sept 13–16; WateReuse Association: Alexandria, Virginia.

de Haas, D.; Turl, P.; Hertle, C. (2004) Magnetic Island Water Reclamation Plant—Membrane Bioreactor Nutrient Removal Technology One Year On. *Proceedings of the IWA World Water Congress and Exhibition*; Marrakech, Morocco, Sept 19–24; International Water Association: London, U.K.

Dodson, D. (2008) Operations and Lessons Learned: Phosphorus Removal from an MBR Treatment Facility (Cauley Creek WRF). *Proceedings of the 2008 South Carolina Environmental Conference*; Mrytle Beach, South Carolina, March 29–April 2; South Carolina Section American Water Works Association/ Water Environment Association of South Carolina: Lexington, South Carolina.

Engelhardt, N.; Linder, W. (2006) Experiences with the World's Largest Municipal Wastewater Treatment Plant Using Membrane Technology. *Water Pract. Technol.*, **1** (4).

Evenblij, H.; van Nieuwenhuuzen, A.; Mulder, J. W. (2007) Hybrid MBR—The Perfect Upgrade? *Water21*, (April).

Fane, A. (2002) Membrane Bioreactors: Design & Operational Options. *Filtr. Sep.*, **39**, 26–29.

GE Power & Water Water & Process Technologies. Membrane Bioreactor (MBR) Design Considerations. http://www.gewater.com/products/equipment/mf_uf_mbr/mbr/design_considerations.jsp (accessed Jan 2011).

Giesen, A.; van Bentem, A.; Gademan, G.; Erwee, H. (2008) Lessons Learned in Facility Design Tendering and Operation of MBRs for Municipal and Industrial Wastewater Treatment. *Proceedings of Water Industry of Southern Africa (WISA) 2008 Biannual Conference and Exhibition;* Sun City, North West Province, South Africa, May 18–22; Water Institute of Southern Africa: Midrand, South Africa.

Guo, W. S.; Vigneswaran, S.; Ngo, H. H.; Xing, W. (2008) Comparison of Membrane Bioreactor Systems in Wastewater Treatment. *Desalination,* **231** (1–3), 61.

Hermanowicz, S. W.; Sanchez-Diaz, E.; Coe, J. (2001) Prospects, Problems and Pitfalls of Urban Water Reuse: A Case Study. *Water Sci. Technol.,* **43** (10), 9.

Itokawa, H.; Thiermig, C.; Pinnekamp, J. (2008) Design and Operating Experiences of Municipal MBRs in Europe. *Water Sci. Technol.,* **58** (12), 2319.

Iversen, V.; Mohaupt, J.; Drews, A.; Kraume, M.; Lesjean, B. (2008) Side Effects of Flux Enhancing Chemicals in Membrane Bioreactors (MBRs): Study on their Biological Toxicity and their Residual Fouling Propensity. *Water Sci. Technol.,* **57**, 117.

Jalla, S.; Daily, J.; Fry, D. (2005) Comparison of MBR Facility Design Processes. *Proceedings of the 78th Annual Water Environment Federation Technical Exhibition and Conference* [CD-ROM]; Washington, D.C., Oct 29–Nov 2; Water Environment Federation: Alexandria, Virginia; pp 4211–4224.

Judd, S. (2006) *The MBR Book*; Elsevier: Amsterdam; p 325.

Koseoglua, H.; Yigita, N. O.; Iversenb, V.; Drewsb, A.; Kitisa, M.; Lesjean, B.; Kraume, M. (2008) Effects of Several Different Flux Enhancing Chemicals on Filterability and Fouling Reduction of Membrane Bioreactor (MBR) Mixed Liquors. *J. Membrane Sci.,* **320**, 57.

Kraume, M.; Drews, A.; Vocks, M. (2005) *Process Configurations for Biotreatment—MBR Concepts for Biological Nutrients Removal*; Technischen Universität Berlin: Berlin, Germany.

Lesjean, B.; Ferre, V.; Vonghia, E.; Moeslang, H. (2009) Market and Design Considerations of the 37 Larger MBR Plants in Europe. *Desalination Water Treat.,* **6**, 227–233.

Lesjean, B.; Luck, F. *IDS Water White Paper—Assessment of the Membrane Bioreactor Technology and European Market Outlook.*

Mao, H.; Phagoo, D.; Penny, J.; Diamond, J. (2005) Design and Optimization Considerations in Integrated MBR BNR Systems. *Proceedings of Water Jam 2005;* Virginia Beach, Virginia; Virginia Water Environment Association: Potomac Falls, Virginia.

McInnis, A.; Alexander, K.; Schneider, C. (2001) Membrane Bioreactor Survey for Operation Optimization and Operator-Friendly Design. *Proceedings of the 74th Annual Water Environment Federation Technical Exposition and Conference* [CD-ROM]; Atlanta, Georgia, Oct 13–17; Water Environment Federation: Alexandria, Virginia; pp 161–169.

Paranjape, S.; Reardon, R.; Cheatham, J. (2009) Designing a Membrane Bioreactor to Minimize Energy Use While Meeting a Low Nitrogen Limit. *Proceedings of the 82nd Annual Water Environment Federation Technical Exhibition and Conference* [CD-ROM]; Orlando, Florida, Oct 10–14; Water Environment Federation: Alexandria, Virginia.

Penny, J. (2007) Lessons Learned in MBR Design after 10 Years of Operation at the Ganges WWTP. *Proceedings of the BCWWA Annual Conference;* Penticton, British Columbia, Canada, April 21–25; British Columbia Water & Waste Association: Burnaby, British Columbia, Canada.

Pinnekamp, J.; Itokawa, H.; Thiemig, C. (2008) Design and Operation of Membrane Bioreactors in Europe. *Water Sci. Technol.,* **58** (12), 2319.

Seyfried, A.; Dorgeloh, E.; Brands, E.; Ohle, P. (1998) Effects of the Membrane Technology on the Dimensioning of Municipal Wastewater Treatment Plants. *Water Sci. Technol.,* **38** (3), 173.

Voldrich, T. E.; Gordon, L. A. (2008) Practical Aspects of Retrofitting to MBR Technology - Mcfarland Creek WWTP- Geauga County, Ohio. *Proceeds of the Membrane Technology Conference;* Atlanta, Georgia, Jan 27–30; Water Environment Federation: Alexandria, Virginia.

Wallis-Lage, C.; Hemken, B.; Steichen, M.; deBarbadillo, C. (2008) Large or Small, New or Retrofit: MBR Design Considerations Are Key. *Proceedings of Singapore International Water Week;* Singapore, June 23–27.

Wilf, M.; Bartels, C.; Bloxom, D.; Christopher, J.; Festger, A.; Khoo, K.; Frenkel, V.; Hudkins, J.; Muller, J.; Pearce, G.; Reardon, R.; Royce, A. (2010) *The Guidebook to Membrane Technology for Wastewater Reclamation*; Balaban Publishers: L'Aquila, Italy; Chapters 7–10; pp 249–324.

Yoon, S.; Collins, J.; Dave, B.; Koppes, J.; Nishimori, K.; Qin, J.; Livingston, D. (2007) Five-Year Experience with Membrane Performance Enhancers fro Membrane Bioreactors: From Mechanistic Studies to Full-Scale Implementation. *Proceedings of the 80th Annual Water Environment Federation Exhibition and Conference* [CD-ROM]; San Diego, California, Oct 13–17; Water Environment Federation: Alexandria, Virginia.

Chapter 6

Membrane Bioreactor Membrane Equipment Procurement

1.0 OVERVIEW

1.1 Introduction

Membrane bioreactor (MBR) membrane equipment represents a large component of a wastewater treatment plant. Membrane bioreactor membrane equipment makes up a substantial portion of construction and operation costs and affects the overall plant footprint, layout, and structural design. Membrane bioreactor membrane equipment currently available is proprietary and operates as a system rather than a series of individual components. The complexities of bidding projects using MBR technology together with the financial investment required have moved the industry toward using alternatives to design-bid-build procurement to evaluate and select MBR membrane equipment. The purpose of this chapter is to provide an overview of current MBR procurement practice to help decision makers, engineers, and suppliers understand the issues and various perspectives associated with procuring MBR membrane equipment.

Table 6.1 summarizes key differences between procuring MBR membrane equipment and "standalone" wastewater treatment equipment such as a pump. A lack of standards, limited number of qualified suppliers, and complex selection and performance requirements translate into complicated procurement. The variety of supplier offerings makes it challenging to use a design-bid-build procurement approach, which requires the owner to have the engineer develop construction documents to accommodate a minimum of two suppliers. To address this situation, industry is using a range of options from a design-bid-build approach, which includes multiple design drawings to accommodate multiple suppliers to sole-sourcing a supplier upfront before commencing with engineering design. This chapter discusses procurement practice with an emphasis on MBR system-specific challenges. Approaches for qualifying, evaluating, and selecting MBR membrane equipment are also discussed. How the procurement process interacts with overall project design, construction, commissioning, and operation is also presented. The chapter ends with a brief overview of issues associated with MBR warranties, MBR acceptance testing, and bid alternates.

1.2 Membranes and Membrane Equipment Procurement Approaches

Industry approaches to procure MBR membrane equipment include the following: prepurchase, preselection, sole source, design-bid-build, design-build, and design-build-operate. For a more detailed discussion of equipment procurement

Table 6.1 Key differences between procuring a pump and MBR membrane equipment.

Issue	Wastewater pump	MBR membrane equipment
Standards	Exist	No standards exist. Supplier-proprietary.
Suppliers	Multiple qualified suppliers	Limited number of qualified suppliers.
Selection	Capital cost	Capital cost, life-cycle cost, or scoring system accounting for cost and nonfinancial factors.
Scope	Equipment type with some appurtenances	Membranes, membrane aeration, permeate pumps, valves associated with membranes and pumps, appurtenances, and controls. May also include biological aeration, biological mixing, and fine screening.
Warranties	Product defects: 1-year standard	Product defects: 3 to 5 years. Membrane modules replacement cost and availability guaranteed for up to 20 years. Product performance (e.g., treat specified flows and water quality for stated operations and maintenance cost). May supply overall treatment biological process warranty.
Performance	Straightforward Achieve flowrates at given head conditions	Complicated. Achieve effluent water quality for range of influent water quality while maintaining required life-cycle power and chemical consumption costs.

practice, refer to Water Environment Federation's *Upgrading and Retrofitting Water and Wastewater Treatment Plants* (WEF, 2005). In addition, for a more detailed discussion of design-build and design-build-operate, refer to Design Build Institute of America's *Design Build Manual of Practice* (Design Build Institute of America, 2009).

Deciding what approach to take should be an owner's decision, with the overall goal being to select an approach that is consistent with the owner's stated preferences around involvement in selection and selection criteria together with project risk associated with overall project cost, schedule, performance, regulatory issues, and funding constraints. Critical to the success of an MBR project is an upfront discussion between the design engineer and the owner on the owner's project goals and preferences in the context of available procurement options to ensure that these goals and preferences are addressed.

1.3 Membrane Equipment Supplier Scope of Services

Scope of supply and scope of services need to be considered early on in the procurement process for MBR membrane equipment. Supplier scope of supply is influenced by many factors including project size, project complexity, and owner preference. Figure 6.1 shows a schematic of how the MBR process affects other unit processes,

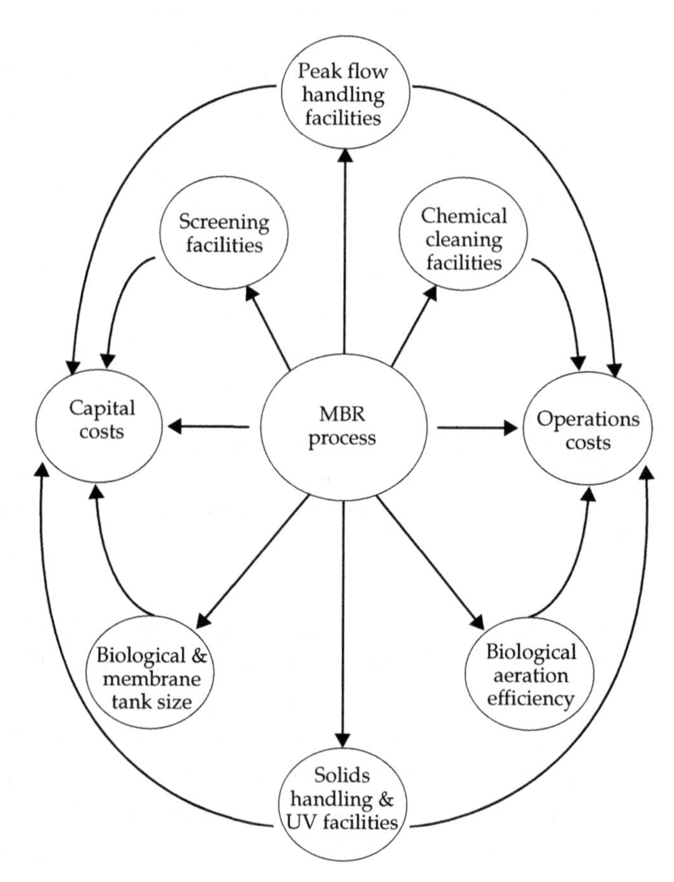

FIGURE 6.1 Effect of MBR process on other unit processes and project capital and operations cost.

which, in turn, influence capital and operations costs. These complex interactions can influence an owner's preference regarding membrane supplier scope of supply, which can range from membranes, membrane aeration, permeate pumps, and membrane controls to all the equipment associated with unit processes shown in Figure 6.1. Supplier scope of supply will also be affected by overall equipment costs; in addition, some owners may want to have only one entity to interact with regarding their significant equipment. In addition to providing equipment, MBR membrane supplier's scope of services may include engineering support services during design, construction, and system startup phases of a project.

2.0 MEMBRANE BIOREACTOR MEMBRANE EQUIPMENT PROCUREMENT

2.1 Introduction

This section presents a general discussion of key considerations and steps that may be performed in support of MBR membrane equipment procurement, focusing on those aspects of procurement that are unique to MBR systems. A discussion of MBR membrane equipment procurement approaches is presented, emphasizing key differences between these approaches and factors that may influence the selection of a procurement method for a particular project. The section ends with a summary comparison of MBR membrane equipment procurement approaches.

2.2 Elements Common to Membrane Bioreactor Membrane Equipment Procurement Approaches

2.2.1 Upfront Preliminary Engineering

Preliminary engineering is necessary to develop a design and conceptual process that forms the basis for a request for proposals or specifications for membrane equipment. Engineering aspects that are typically required prior to issuing formal solicitation for membrane equipment include definition of design flows, treatment objectives, preliminary facility layout, and redundancy requirements. Because there is no standardization among MBR membrane systems, many of the design details are unique and specific to each membrane manufacturer. This poses an additional complexity when conducting upfront engineering prior to the selection of a particular membrane technology because the preliminary design may need to accommodate more than one membrane product. In practice, this may be managed by conducting the preliminary design based on a selected membrane vendor and allowing for variations required for alternative equipment.

2.2.2 *Selection Process*

The process of selecting a membrane equipment vendor may follow a number of different paths, depending on the size and complexity of the project, contractual arrangements, and preferences of the owner. Options for selecting membrane equipment may include the following:

- Statement of qualification,
- Request for proposal (RFP), and
- Statement of qualifications followed by RFPs.

2.2.3 *Contract Development*

The structure of a contract for the procurement of membrane equipment is largely determined by the overall contractual arrangement for a project. For example, the membrane equipment supplier could be contractually engaged directly with the owner, with a general contractor, with a design-build engineer, or with another party involved in the project. The contract may also be split into a portion for engineering services to develop and refine the plant design, followed by the contract for equipment procurement. In the case of membrane filtration equipment, it is important to consider the contractual arrangements for the life of a project, including the contractual relationship for extended membrane warranties after completion of construction and startup of the plant. As with the purchase of any process equipment, consideration should be given to including detailed design and performance criteria in the procurement contract.

2.2.4 *Bid Evaluation, Award, and Negotiation*

An important aspect of the development of an RFP or specification is defining the bid evaluation criteria and method. Membrane equipment systems are typically evaluated based on both cost and noncost factors, where the specific criteria are defined based on the best interests and priorities of the owner. Table 6.2 summarizes criteria that may be considered in an MBR membrane equipment evaluation.

2.2.5 *Coordination during Final Design and Construction*

It is not uncommon for an MBR project to experience changes in design or scope after a membrane equipment contract has been awarded because equipment selection

TABLE 6.2 Typical MBR membrane equipment evaluation criteria.

Cost factors	Non-cost factors
CAPEX (capital cost)	Vendor experience/references
OPEX (operating cost)	Equipment operability/maintainability
Membrane replacement	Vendor service offering
Energy consumption	Vendor corporate financial stability
Cleaning chemical use	
Life-cycle cost	

is typically based on preliminary engineering only. Issues that may be considered when an owner or design engineer reviews design changes may include potential cost savings, effects to project schedule, potential operating cost savings, and so on. In instances where design changes result in a change to the scope or requirements of membrane equipment, the owner may elect to revise and/or conform the equipment procurement contract to reflect the final design and equipment provided.

2.3 Membrane Bioreactor Procurement Approaches

Table 6.3 summarizes the steps associated with various MBR procurement approaches. For purposes of discussion, these approaches are treated as if they are all mutually exclusive. This being said, a design-bid-build overall project procurement may be coupled with prepurchase, preselection, or sole source for membrane equipment procurement. Referring to Table 6.3, the approaches can be compared as follows:

- Project schedule
  - Sole source tends to require the shortest project completion time.
  - Design-bid-build tends to require the longest project completion time.
- Project cost
  - Design-build-operate establishes a guaranteed project life-cycle cost (for specified life-cycle duration).
  - Prepurchase and Preselection make the membrane equipment-related costs transparent.
- Project complexity
  - Sole source tends to be the least complicated project approach.

TABLE 6.3 Membrane bioreactor procurement approaches.

Approach	Description
Prepurchase	Owner develops contract documents and solicits supplier bids (request for qualification [RFQs]/RFPs or RFP). Owner can be involved in determining selection criteria/evaluation process. Owner selects supplier based on competitive bid evaluation. Owner purchases membrane equipment before final engineering begins. Engineer may work with supplier to develop final design and construction documents around purchased equipment. Equipment delivered to owner and furnished to Contractor to minimize overall project construction schedule.
Preselection	Owner develops RFQs/RFPs or RFP and supporting contract documents to solicit supplier bids. Owner can be involved in determining selection criteria/evaluation process. Owner selects supplier based on competitive bid evaluation. Owner compensates supplier for engineering services during design (if this service is included in the contract). Engineer works with supplier to develop final design and construction documents around selected equipment. Contractors directed to carry equipment cost in their project bids. Equipment contract may be assigned to selected general contractor during project construction.
Sole source	Owner negotiates price with equipment supplier to sole-source supplier. Engineer may work with supplier to develop preliminary and final design and construction documents around sole-sourced equipment. Equipment delivered to owner and furnished to contractor to minimize overall project schedule.
Design-bid-build	Owner directs engineer to develop contract documents around at least two suppliers. General contractor selects which membrane supplier to include in their bid, based on equipment and construction-related membrane costs. Owner selects general contractor based on lowest project bid.
Design-build	Owner develops RFQs/RFPs or RFP. General contractor selects which membrane supplier to include in their bid, based on equipment and construction-related membrane costs. Owner selects general contractor based on lowest project bid (if quality-based selection is the approach taken, different selection criteria).

(continued)

Table 6.3 Membrane bioreactor procurement approaches (*continued*).

Approach	Description
Design-build-operate	Owner develops RFQs/RFPs or RFP. Selection documents are complex given long-term nature of design-build-operate. General contractor selects which membrane supplier to include in their bid based on lowest life-cycle cost (equipment, construction-related, and operations). Owner selects general contractor based on lowest life-cycle project bid.

- Design-build-operate tends to be the most complicated project approach given the complexities of the contract.
- Both prepurchase and preselection are complicated and require an owner's attention to multiple contracts as part of the overall project.

- Project risk
  - Design-build-operate mitigates cost-related risks by establishing life-cycle cost.
  - Sole source mitigates schedule-related risks by decreasing engineering time and equipment delivery time to the contractor.
  - Design-build-operate mitigates performance risks by establishing performance goals tied to financial incentives.
  - Both prepurchase and preselection can mitigate membrane performance risks depending on the criteria used for selection and whether or not a supplier process warranty is included.

Table 6.4 compares MBR procurement approaches. It is important to note that only two of the approaches (sole source and prepurchase) require owner equipment financial commitment before construction begins. All the approaches in Table 6.4 have either direct or indirect competitive bidding built in, except for sole source. In addition, prepurchase, preselection, and sole source provide an owner with the ability to consider non-financial-related criteria when selecting membrane equipment, if this is an important preference. Finally, it is important to note the various contracting entities associated with both equipment and project procurement. When multiple contracts are in place, the owner needs to make sure there is coordination between

Table 6.4 Comparison of MBR procurement approaches.

Issues	Equipment procurement approaches					
	Prepurchase	**Preselection**	**Sole source**	**Design-bid-build**	**Design-build**	**Design-build-operate**
RFQs/RFPs or RFP developed for owner	X	X		X	X	X
Owner influences equipment selection criteria	X	X	X			
Equipment competitively bid	X	X		X[a]	X[a]	X[c]
Equipment contract entities Owner-Equipment supplier	X	X	X			
Equipment contract entities Owner-General contractor				X	X	X
Equipment contract entities General contractor-Equipment supplier		X[b]		X	X	X
Owner pays for equipment before construction begins	X		X			
Guaranteed maximum project construction price					X	X
Guaranteed maximum project operations price over specified period						X

[a] General contractor selects supplier with combination of lowest equipment cost and related construction cost.

[b] Owner may assign equipment contractor to general contractor during construction.

[c] General contractor selects supplier with combination of lowest equipment, construction, and operations cost.

these contracts throughout the project life cycle to ensure project delivery will meet stated project requirements.

In summary, there are advantages and disadvantages to all of the approaches discussed. In addition, an owner may not have the ability to sole source because of the limits to their ability to do so over a certain cost and/or because of federal or state funding requirements or laws (e.g., U.S. Department of Agriculture Rural Development funding requirement of two or more specified equipment manufacturers). The owner may also have concerns about entering into a design-build-operate procurement because of the complicated nature of these types of contracts. As discussed earlier, the key to successful procurement is to work closely with the owner to understand their preferences and priorities before beginning the procurement process, which will dictate the types of procurement approaches to be considered.

3.0 OTHER TOPICS TO CONSIDER WHEN PROCURING MEMBRANE BIOREACTOR MEMBRANE EQUIPMENT

3.1 Membrane Bioreactor Warranties

Warranties provided by suppliers of MBR membrane equipment for MBR projects range from equipment warranties, which include membrane warranties, to an overall biological process warranty. Warranty duration and specified performance goals translate into additional costs to the owner that shift risk to the MBR membrane equipment supplier. Part of the overall evaluation of which warranties to include in project documents is to present the range of options to the owner and discuss the merits and cost tradeoffs associated with these types of warranties. General warranty issues to address with the owner are as follows:

- Warranty terms and conditions need to be clearly specified and need to include limitations in the warranty;
- Warranties should be provided by the equipment supplier, not the equipment representative;
- Warranties can be either nonprorated or prorated. This issue typically comes up with membrane module and element replacement;
- Maintaining a complete record of service dates for all MBR system equipment to maintain the terms and conditions of the warranty; and
- Warranties may require the owner to conduct regular sampling and testing of influent and effluent quality and document predefined operating parameters.

Membrane bioreactor system warranties can be classified as follows: a membrane system equipment warranty, a membrane module or element warranty, an

MBR system *performance* warranty, and an MBR system *biological process* warranty. A membrane system equipment warranty typically covers defects and workmanship, with failed equipment or components replaced by the supplier at no cost to the owner. A typical duration for this type of warranty is 1 or 2 years, beginning from a significant project milestone like successful completion of acceptance testing. Limitations to this type of warranty may include physical damage after equipment acceptance, faulty installation not attributable to the supplier, alteration of equipment and/or controls without prior authorization from the supplier, and lack of or improper maintenance as specified in the supplier's operations and maintenance manual.

Membrane module and element warranties are important for the owner given the proprietary nature of membranes and changing technology. This type of warranty typically specifies membrane module and element cost (tied to a metric-like inflation or an Engineering News Record index) and availability (up to 20 years from a significant project milestone like successful completion of acceptance testing). The warranty could be structured to allow the supplier to provide a changed or newer membrane module or element, with the stipulation that the owner would bear no additional cost to accommodate this type of replacement. Warranty periods for defects, workmanship, and normal wear for a membrane module or element can range from 5 to 8 years for full-replacement cost to up to 10 years on a prorated basis, depending on whether or not the owner requires extended warranty coverage. This type of warranty can include statements requiring the supplier to identify water quality parameters, instrumentation, and control programming that are required to properly operate and maintain membrane modules or elements, and provide alarms and/or shutdown limits to prevent operation outside of the limits established by the supplier. Limitations to this type of warranty may include membrane exposure to wastewater (e.g., as quantified by mixed liquor suspended solids), process design (e.g., as quantified by solids retention time) or treatment (e.g., polymers) and cleaning chemicals at concentrations above levels acceptable to the supplier as identified in the supplier's original proposal and in the supplier's operations and maintenance manual.

An MBR system *performance* warranty may cover items such as membrane effluent quality, as measured by effluent turbidity, and guaranteed chemical cleaning (type, frequency, and duration) and power usage costs based on the wastewater flow regime (average and peak flows and peak flow duration), plant effluent water quality, and wastewater influent water quality. The warranty period for this type of warranty can range from 1 to 5 years. An MBR system *biological process* warranty states that for the specified flow regime (average and peak flows and peak flow duration) and wastewater influent water quality, the MBR system will achieve the stated plant effluent water quality (e.g., biochemical oxygen demand, total nitrogen, and possibly, total phosphorus). The warranty period for this type of warranty can range from 1 to 10 years. Should these types of warranties not be met, the supplier may be required to provide improvements that could include replacing membrane equipment and/or membrane modules and elements, adding additional membrane equipment and/or membrane modules and elements (if membrane tankage space permits), or paying a prorated amount equal to the insufficient capacity to achieve specified cleaning frequencies. Limitations to these types of warranties include those limitations listed for membrane module and element warranty limitations, plus flow regimes and wastewater influent quality outside of the warranted values.

One of the challenges owners face regarding MBR system performance and MBR system biological process warranties is to determine if the MBR system meets the warranty at plant design conditions, which may not occur for some time given that these conditions typically correspond to a future or buildout condition. For plants with multiple MBR treatment trains, design conditions can be simulated by routing design conditions through a subset of available MBR treatment trains. For this condition, two tests can be conducted to help evaluate MBR system response to design conditions and warranty compliance: (1) stress testing (several weeks, typically conducted during plant acceptance testing); and (2) performance testing (several months, typically conducted during long-term performance testing). These tests are described in the following section.

3.2 Membrane Bioreactor Acceptance Testing

One of the challenges facing operators of any new or improved facility is to determine how the facility will perform at design conditions. Lack of long-term MBR operational experience, complexity, and the integral nature of MBR membranes to overall biological process performance makes testing a key part of the overall approach of ensuring plant performance. Types of acceptance testing include the following:

- Clean water test,
- Thirty-day plant acceptance test,

- Stress test, and
- Long-term system performance test.

Clean water testing is conducted prior to testing with wastewater to ensure proper system operation and test system hydraulics based on system operational design criteria. Plant acceptance testing is typically conducted as part of plant startup with wastewater to demonstrate that the MBR system meets performance requirements. As previously discussed, plants are typically designed with future design capacity in mind and, therefore, do not currently experience flows corresponding to the design condition. One approach for addressing this situation is to conduct a stress test, whereby flows are treated using only a subset of available MBR treatment trains to demonstrate the system's ability to perform under design flow conditions (both average and peak) at the MBR membrane equipment supplier's proposed net flux design. Stress testing typically consists of a week of average flow conditions, followed by loading the MBR treatment train at the peak flow event for the required duration, followed by membrane cleaning. This two-step stress test can be conducted several times during plant acceptance testing, collecting the data that allow for calculating net flux, permeability and trans-membrane pressures. These results provide the basis for whether or not the owner accepts the MBR membrane equipment supplier's flux design. Failure to achieve stress test criteria (e.g., clean membrane permeability and transmembrane pressure [TMP] values established at the beginning of stress test no. 1 and at the end of stress test no. 2 shall be within 10%) would result in the owner's rejection of the MBR membrane equipment supplier's membrane flux design and result in the basis for a warranty claim.

Long-term performance testing may be conducted over a period ranging from 6 to 24 months to further confirm MBR system operation and performance. During this period, flows can be directed to a subset of available MBR treatment trains to simulate design conditions over a duration longer than the aforementioned stress test, providing further confirmation of MBR system performance. During both acceptance and long-term system performance testing, the types of data collected to confirm that the MBR system is meeting performance requirements may include MBR system production, permeate water quality, flux rates, permeability, TMPs, backwash and backpulse frequency (if required), chemically enhanced backwash, clean frequency (if required), and recovery chemical clean frequency. In addition, the current unit chemical and power costs at the time of the long-term system performance test can be used to calculate chemical and power costs.

3.3 Bid Alternates

Bid alternates require careful, upfront consideration when procuring MBR membrane equipment. If given the opportunity, MBR membrane equipment suppliers will offer additional equipment as part of a bid alternate to their baseline scope of supply. Membrane bioreactor membrane equipment suppliers maintain business relationships with original equipment manufacturers, which allow MBR membrane equipment suppliers to purchase equipment at a cheaper price than could be negotiated with a general contractor. The benefits of this approach for the owner are having fewer entities to interact with regarding equipment performance issues and the potential that a bid alternate provides the owner a lower equipment cost relative to similar equipment that would be procured under the general contractor's contract. The challenge associated with bid alternates is to clearly define their scope of service (as opposed to an open-ended, supplier-specified scope of service) and develop a contractual approach that will make the cost of the package equipment more transparent to the owner so they can evaluate the value in having MBR membrane equipment suppliers "package" additional equipment as part of overall procurement. One approach for achieving this goal is to call out all the package equipment items as unit prices in the bid form. Using this approach, the owner could see the cost of individual equipment components and, if desired, check with component equipment suppliers other than those suppliers who provide equipment to the MBR membrane equipment package provider. This would allow for confirming component equipment costs and determining whether there is any potential cost savings associated with equipment packaging.

4.0 REFERENCES

Design Build Institute of America (2009) *Design Build Manual of Practice*, 2nd ed.; Design Build Institute of America: Washington, D.C.

Water Environment Federation (2005) *Upgrading and Retrofitting Water and Wastewater Treatment Plants;* WEF Manual of Practice No. 28; Water Environment Federation: Alexandria, Virginia.

Chapter 7

Membrane Bioreactor Operation

(continued)

Successful operation of any treatment system involves the proper operation of each of its parts. This is even more critical for a membrane bioreactor (MBR) system given the interrelationships between the various unit processes. For example, as shown in previous chapters, the performance of the activated sludge system can directly affect the performance of the membrane system. Similarly, the performance of upstream pretreatment processes (i.e., fine screens, equalization) affects the performance of activated sludge and membrane systems. Therefore, although an MBR system is capable of producing a consistent, low total suspended solids (TSS) effluent, it is critical to ensure proper operation and maintenance (O&M) procedures are followed for both the membrane process and all of the upstream unit processes. It should be noted that the material covered in this chapter focuses on immersed membrane systems; however, many of the same principles apply to less common externally located pressurized membrane MBR systems.

1.0 OPERATION AND ROUTINE MONITORING

1.1 Membrane Bioreactor Standard Procedure

An MBR typically includes the following: bioreactor (activated sludge); membranes; mixed liquor suspended solids (MLSS); return activated sludge (RAS); waste activated sludge (WAS); two air-supply systems with blowers (one for the activated sludge system and one for the membranes); chemical feed; metering and storage systems; compressed air and online monitoring; and control systems.

1.1.1 *Initial Startup and Commissioning*

The initial startup period is an important step of the MBR installation process and is an essential consideration during facility planning and design process. This phase often includes tank and piping verification, large membrane subunit installation, membrane integrity and performance testing, system diagnostic testing, and operator training.

Prior to installing large membrane subunits, the membrane tank should be hydrostatically tested and cleaned to remove all debris larger than the influent screen openings (if applicable). All tanks, piping, valves, and pumps should be flushed at design flowrates prior to installation of the membranes. All construction debris in the piping and tanks must be removed to ensure the membranes are protected from residual materials (i.e., metal, soil, and dust) during the startup process.

Until installation occurs, the membrane subunits should be stored in a dry environment away from direct sunlight and extreme temperatures (i.e., freezing temperatures). Storage and handling requirements can be specific to the type of membrane materials and manufacturer and it is important to understand these requirements early and make necessary arrangements. It is also important to ensure that the area where the membrane subunits will be unpackaged is level, dry, and free of debris.

Small membrane subunits are typically shipped with a liquid preservation solution to prevent the membranes from drying out and to control microbial growth. The preservation fluid (often glycerin) should be disposed of per manufacturer's recommendations, but can be "flushed to waste" after the membranes are installed (note that excessive foaming may occur). For some suppliers, preservation fluid can be biodegraded in the initial stages of bioreactor seeding and may not require flushing to waste. However, the high chemical oxygen demand (COD) of some preservation fluids (i.e., glycerin) and their effects on the wastewater treatment process should be considered before disposal. Connections for membrane subunits should also be verified at this time and membrane integrity confirmed, typically with an air test.

1.1.1.1 Clean Water Testing Once the large membrane subunits have been installed, and proper operation of the mechanical components has been confirmed (i.e., piping integrity, tank or membrane subunit support frame deflection, distortion, or leakage), performance testing of the membrane should be verified through Clean Water Testing. Clean Water Testing is performed to confirm that the membrane performance is as stated and to establish a baseline or benchmark with respect to the membrane's flux and permeability. Over time, both flux and permeability will decrease because of irreversible fouling, confirming that this decrease is within acceptable range and can be accomplished by comparing the results of future Clean Water Testing with these baseline results.

When reviewing flux and permeability, it is important to be cognizant of the effect of temperature and the different types of flux (i.e., instantaneous, net, average). For a more detailed discussion of these terms, refer to Chapter 2 and the glossary in Appendix B.

Clean Water Testing typically involves the following procedures:

1. Filling the membrane tanks with clean water—clean water sources could be tap water, non-potable water, or groundwater; confirm the acceptability of the clean water source with the membrane manufacturer otherwise the membranes could be damaged;
2. Starting the air scour system and confirm uniform air supply to the membrane subunits;
3. Inspecting the membranes—check for items that are clearly different from subunit to subunit and for any apparent damage to the large and small membrane subunits;
4. Generally verifying proper operation of the mechanical components;
5. Starting permeation and documenting the permeate flowrate, transmembrane pressure (TMP), and water temperature for multiple operating points over a wide range of flows (i.e., 25 to 125% of the design flow); and
6. Instrumentation and control verification, calibration, establishing setpoints, and alarms checked.

The duration of the clean water test should be kept to a minimum as it tends to cause membrane fouling from construction dust. Chapter 6 contains additional information on Clean Water Testing.

1.1.1.2 System Seeding After the clean water test and all upstream and downstream processes have been tested, suitable seed biomass (such as RAS or WAS) can be introduced to the system. Ideally, the seeding and startup phases will commence immediately after completion of Clean Water Testing; this will reduce the potential for damage to the membrane system from ongoing construction activities and minimize the possible drying out of the membranes. During the seeding phase, aeration equipment must be engaged in the bioreactor and/or membrane tank for the biological treatment process; it is important to note that membrane filtration of raw wastewater may cause membrane fouling. It is recommended that the seed biomass be procured from a treatment process of the same type of wastewater and the seed biomass screened prior to introduction to the bioreactors. While feeding raw wastewater, aeration systems (bioreactor and/or membrane tank) should be operating to maintain aerobic conditions and recirculation pumping (along with mixers, if applicable) should be operating to distribute seed sludge throughout the system. It should be noted that, depending on initial plant flows, the bioreactor may be operated at lower MLSS concentrations than designed.

1.1.2 *Typical Operation*

In general, raw wastewater quality and operating conditions affect overall MBR performance. To maintain stable operating conditions, operators should measure and record various parameters. These operational parameters often include the following:

- Air scouring flowrate;
- Blower discharge pressure and amperage;
- Transmembrane pressure;
- Raw wastewater characteristics (i.e., alkalinity, temperature, pH, 5-day biochemical oxygen demand [BOD_5], COD, TSS, total Kjeldahl nitrogen, total phosphorus);
- Bioreactor and membrane tank water quality (i.e., alkalinity, temperature, pH, MLSS, dissolved oxygen, solids retention time [SRT], food-to-microorganism [F/M] ratio, viscosity);
- Waste activated sludge/RAS flowrates;
- Permeate water quality (i.e., alkalinity, temperature, pH, BOD_5, COD, dissolved oxygen, ammonia (NH_3), total nitrogen, total phosphorus, TSS, turbidity); and
- Permeate flowrate.

Specific water quality parameters that should be monitored vary with the characteristics of the influent wastewater and treated water quality discharge requirements. Monitoring the process of an MBR system is similar to that required for a conventional activated sludge (CAS) system (for more information on CAS systems, refer to

Operation of Municipal Wastewater Treatment Plants [WEF, 2007]). However, there are a few key areas important to MBR systems that should be monitored. These include membrane hydraulic performance, membrane integrity, membrane mechanical testing, and pretreatment.

Membranes should be monitored daily for fouling, deterioration, and aging. Flow meters and pressure gauges allow for calculation of the flux and TMP, which both provide valuable information. The flux and TMP can then, in turn, be used to calculate permeability (refer to Chapters 1 and 2 and Appendix B for definitions and typical ranges). Using this information, operations staff can make informed decisions on how many membrane trains should be in service, when to initiate chemical cleaning events, and when membranes may have reached the end of their useful life.

In addition, two other tests are useful at providing insight to how the activated sludge system may affect membrane performance and cleaning frequencies. The colloidal total organic carbon (TOC) fraction in MLSS can be measured and provide an indication as to membrane fouling potential (particularly pore blocking) of the activated sludge (i.e., higher colloidal fraction increases the fouling potential).

The time-to-filter (TTF) test is a straightforward test in which the time it takes to filter a specific volume of MLSS is measured to determine its "filterability." Typically, a desirable TTF is between 50 and 200 seconds; although, a fair TTF is between 200 and 300 seconds and difficult-to-filter sludges have TTFs longer than 300 seconds. High membrane TMPs are often associated with membrane fouling; however, high TMPs coupled with high TTF results are more likely an indicator of poor performance by the activated sludge system. The TTF test can provide similar feedback for MBRs as the sludge volume index (SVI) test does for secondary clarifiers.

In addition to monitoring parameters in the permeate such as BOD_5, TSS, and ammonia, continuous measurement of permeate turbidity is used to monitor membrane integrity; an air test (bubble test, pressure decay test) can also be used to verify the integrity of the membranes, although this requires removing the membranes from service and, in some instances, removing them from the membrane tank (see Chapter 4).

The membrane air scour system should be checked on a daily basis to ensure that air is being supplied and distributed evenly across the large membrane subunit(s). Non-uniform air distribution may cause membrane clogging or indicate a potential failure in the aeration system.

The liquid level in the membrane tank should be monitored to confirm that the MBR is operating in the appropriate range. If this condition is not met, the operator should check to see that the liquid level instrument is functioning properly, review the condition of the permeate pump(s) and TMP, and make necessary adjustments to the system to correct the problem.

As has been mentioned in this manual, pretreatment, particularly screenings equipment, is important in an MBR. One method for confirming proper operation of the screens is the sieve test. The sieve test involves a known volume of representative MLSS through a stack of sieves with appropriate openings (i.e., 1 to 3 mm), weighing the collected solids, and comparing the results (typically in mg/L) to baseline numbers. The results of this test are used to confirm the effectiveness of upstream pretreatment systems (i.e., screens, grit, primary clarifiers) and ensure that bypassing is not occurring.

1.1.2.1 Permeation Mode Most of the time, membranes will be operating in permeation mode. The permeate flowrate is typically based on the influent flowrate to the activated sludge system and is modulated using variable frequency drives on permeate pumps (or permeate flow control valves for a gravity permeate system). Permeate passes through the large membrane subunit and is collected in a manifold (or header pipe) prior to being discharged to downstream processes. The flow through the membranes is not allowed to reach a point above the maximum allowable TMP and should either automatically shut down and/or begin a cleaning cycle. Solids buildup at the surface of the membrane is controlled by the membrane air scour system, which may operate either intermittently (25 to 50% of the time) or continuously (100% of the time).

1.1.2.2 Relaxation Mode During relaxation, permeation is suspended while the air scour remains in operation. Despite the efforts of air scour to control buildup of solids at the membrane surface during permeation, some solids will adhere and possibly enter the membrane. By suspending permeation, solids become dislodged or "released" from the membrane surface and removed by the air scour. When permeation resumes, there is typically a decrease in TMP required to produce a certain permeate flowrate compared to the same flowrate prior to relaxation. Typical relaxation-permeation cycles are 30 to 120 seconds relaxation for every 5 to 15 minutes of permeation. For further detailed information on relaxation cycles, refer to Chapter 4, Section 5.0.

1.1.2.3 Backwash Mode Occasionally, relaxation of the membranes is not adequate for proper fouling control and backwashing of the membranes is beneficial. Backwashing (also referred to as *backpulsing*) involves reversing the flow through the membranes to remove solids particles from the membrane pores using permeate. The permeate is pumped through the membranes at low pressure and high flowrates (up to 150% of normal permeation flowrate) to remove particles that may have adhered to the membrane surface. Although it was common practice to backwash membranes at the end of each permeation cycle in the past, current cleaning strategies use backwashing less frequently. Operations staff may also elect to perform a chemically enhanced backwash, where a chemical solution is injected in the backwash feed stream prior to the membranes. Typical cleaning solutions consist of dilute concentrations of sodium hypochlorite (i.e., bleach) and citric acid; however, a variety of other cleaning chemicals are commercially available for various fouling types (see Section 6.6.4 of Chapter 4 for further information on chemical cleaning).

1.1.3 *Membrane Tanks in Standby Mode*

Diurnal and seasonal flow patterns may allow the removal of a membrane tank from service temporarily (less than 1 day), placing it in standby mode. In general, this will entail stopping permeation and reducing the air scour rate (i.e., 1-minute operation every 30 minutes) while maintaining flow of MLSS through the membrane tank. For periods longer than 1 day but less than 2 weeks, the tank should be drained and refilled with clean water and a disinfectant. For longer periods, the membranes will likely need to be removed from the tanks and stored using a preservative.

1.1.4 *Shutdown of the Membrane Tanks*

Occasionally, it is necessary to shut down a membrane for preventative maintenance of the membrane subunits, the membrane tank, or other ancillary systems. Although the procedures may differ slightly between various membrane manufacturers, there is common agreement that once membranes have been installed in a tank, they should be kept wet at all times; however, the membranes can typically be allowed to stand "in the dry" for a few hours, depending on local weather conditions. In general, specific shutdown methods, duration, and membrane preservation requirements should always be discussed with the membrane supplier on a case-by-case basis to maintain a valid warranty. Additional information regarding membrane storage, handling, and replacement is discussed in Section 2.0 of this chapter.

1.1.5 *Safety*

Safety is always an important consideration when operating any wastewater system and an MBR is no different. A detailed discussion of industry safety requirements is outside the scope of this MOP; however, a good starting point is to consult regulations published by the Occupational Health and Safety Administration and safety regulations and procedures developed by the organization responsible for operating the MBR system. Areas of concern that are specific to MBRs include

- Accessing areas around the tops of membrane basins (immersed systems),
- Entry into the membrane basins (immersed systems),
- Use of overhead cranes (varies with installation and design),
- Overpressurizing membrane vessels (pressurized systems),
- Chemical storage and handling, and
- Noise from membrane air scour blowers (varies with installation and design).

1.2 Membrane Bioreactor Process Effects

1.2.1 Upstream Effects on the Membrane Bioreactor

There are several upstream preliminary and primary treatment processes that can potentially affect the MBR system, including screenings and grit removal, equalization, and primary clarification. Upstream treatment considerations are discussed in more detail in Chapter 4, Section 3.0.

Traditionally, MBR systems have been designed to operate at much higher MLSS concentrations than conventional biological treatment processes. This type of operation results in high RAS flowrates (often 4 to 5 times the influent flowrate) with high BOD_5 and COD removal efficiency.

In an MBR system, membrane fouling is directly determined by the mixed liquor biomass characteristics and hydrodynamic conditions (i.e., aeration intensity, bubble size, MLSS concentration, viscosity). The bioreactor operating conditions (i.e., hydraulic retention time [HRT], SRT, F/M), coupled with feedwater characteristics have indirect relationships on membrane fouling by altering biomass characteristics.

Literature suggests that factors affecting membrane fouling can be classified in the following four groups: membrane materials, biomass characteristics, feed characteristics, and operating conditions (Meng et al., 2009). Interactions between the groups are complex and beyond the scope of this chapter (see Chapter 2 and the references for more indepth discussions). However, it is important to note some of the relationships that exist between various fouling factors and membrane fouling potential, as shown in Table 7.1.

This list is not intended to be all inclusive and the reader is encouraged to review literature and other references for more detailed information on specific parameter(s)

TABLE 7.1 Relationship between various operating parameters and their effect on membrane fouling (adapted from Meng et al. [2009]).

Parameter	Effect on membrane fouling
MLSS concentration ↑	Fouling potential ↑
Filamentous bacteria ↑	Viscosity ↑
Temperature ↓	Viscosity ↑
Viscosity ↑	Membrane permeability ↓
Air scour intensity ↑	Membrane permeability ↑
Extracellular polymeric substances (EPS) ↑	Fouling rate ↑
Soluble microbial products (SMP) ↑	Fouling potential ↑
SRT ↑ or SRT ↓*	Fouling potential ↑
HRT ↓	Fouling potential ↑

Notes:

↑ or ↓denotes an increase or decrease in the parameter, respectively.

* Operating outside of the preferred SRT range will increase the fouling potential.

of interest. Biological fouling effects are discussed in more detail in previous chapters of this manual of practice.

While a significant contributor to membrane fouling in an MBR is the solids content of the mixed liquor, there are other potential foulants. These include oil and grease, coagulants (alum, ferric chloride, polymers, etc.), and antifoaming chemicals (compatibility should be confirmed with the MBR manufacturer). Some of those foulants cannot be removed from the membrane surface, causing irreversible loss of membrane capacity. Some of these potential foulants are present in the incoming wastewater and will be removed by properly operating preliminary treatment processes, while some are added as chemicals to the treatment stream as part of the treatment process. Careful consideration of where and how the chemicals will be used must be undertaken during design, with full input from the MBR supplier.

1.2.2 *Membrane Bioreactor Effects on Downstream Processes*

More detailed discussions of MBR effects on downstream processes are included in previous chapters, but it is important to mention them here as they relate to operations.

The nature of MBR effluent affects downstream processes in several ways because of its consistently low solids concentration (<1 mg/L TSS). This provides a reliable feed that eliminates some of the variability and, therefore, allows for smoother operation of these processes. This eases efforts related to planning process operating strategies and estimating preventive maintenance schedules.

Hydraulically, unless there is substantial flow equalization upstream of the MBR process, the volume of effluent will vary with diurnal fluctuations, not unlike the effluent from a conventional treatment system. However, because of the intermittent relaxation and/or backwash cycle when a portion of the membranes are not permeating, brief periods of lower flow are possible on a frequent basis. This phenomenon should be reviewed and addressed during the design phase and may, in some instances, require a "break" tank immediately downstream of the membrane process. This phenomenon is much less prevalent as the capacity of the treatment plant increases as an increasingly smaller percentage of the membranes are not permeating at any given time because there are typically a greater number of membrane trains in operation. Some downstream processes include advanced and tertiary filtration, disinfection, and solids and sludge handling.

Advanced filtration processes can include denitrification filtration, nanofiltration, and reverse osmosis. All of these processes benefit from low solids concentration in the feed stream, whereas nanofiltration and reverse osmosis typically require pretreatment by either microfiltration or ultrafiltration membranes. The microfiltration or ultrafiltration membranes in the MBR system serve this purpose while also serving as the final clarification step for the secondary treatment process. It should be noted, however, that because of the high solids concentration in the MBR system, a breach in the microfiltration and ultrafiltration membrane could potentially result in significant fouling of downstream membrane processes. Conversely, in CAS systems, a breach in the microfiltration and ultrafiltration membrane is potentially less problematic because they typically follow secondary clarifiers and/or tertiary media filters. Managing the flow from the MBR is important to ensure adequate and consistent flow to membrane processes. Although a membrane feed tank is typically included in the treatment process, operations staff must be aware of flow requirements, particularly when removing MBR membrane basins from service and during cleaning events.

Disinfection processes commonly include UV or chlorine. These processes also benefit from the low solids concentration in the MBR effluent reducing UV and chlorine dosages. As with advanced and tertiary filtration processes, care must be taken to ensure adequate and consistent flow from the MBR, particularly for UV disinfection.

Effects on UV disinfection are covered previously in this MOP. Advanced oxidation process (AOP) is also becoming more popular in combining UV with ozone or hydrogen peroxide for enhanced trace organics removal. For disinfectant residual in the distribution system following AOP, it would still be appropriate to add chlorine.

The amount of WAS generated by an MBR plant is comparable to the amount generated by a CAS plant, assuming a similar SRT. However, the higher concentration of MBR WAS obviously affects how the WAS is handled and pumped and must be considered when evaluating thickening, dewatering, and digestion processes. The composition of MBR WAS is also inherently different from CAS WAS because of the bacterial speciation and higher incidence of smaller particles. However, there is currently no evidence to support the conclusion that MBR sludge is more difficult to dewater than conventional sludge.

2.0 MAINTENANCE

Successful performance of any equipment or system can be directly related to proper maintenance. The preventative maintenance schedule suggested by the manufacturer should be used as a starting point to develop a customized maintenance schedule for an MBR system. With the numerous different types and configurations of MBR systems available, it would be difficult to enumerate all of their required preventative maintenance tasks; this section will describe the procedures most common to an MBR system.

2.1 Membrane Cleaning

Periodically, membranes will need a more rigorous cleaning than that provided by the membrane air scour and routine backwash and relaxation and/or backpulse events. Fouling results from the interaction between the membrane and mixed liquor and is often defined as "the reduction of flux (at a given TMP) caused by the buildup of contaminants" (WEF, 2006). Membrane fouling is characterized as

- Physically reversible—foulants loosely attached to the membrane surface and present in the membrane pores; these are typically removed by air scour, but in some instances the sludge can build up to the extent where manual cleaning is necessary. Examples of membrane sludge buildup (which can also be attributed to stringy material) are shown in Figures 7.1 and 7.2. Membrane sludge buildup can typically be attributed to poor performance of upstream

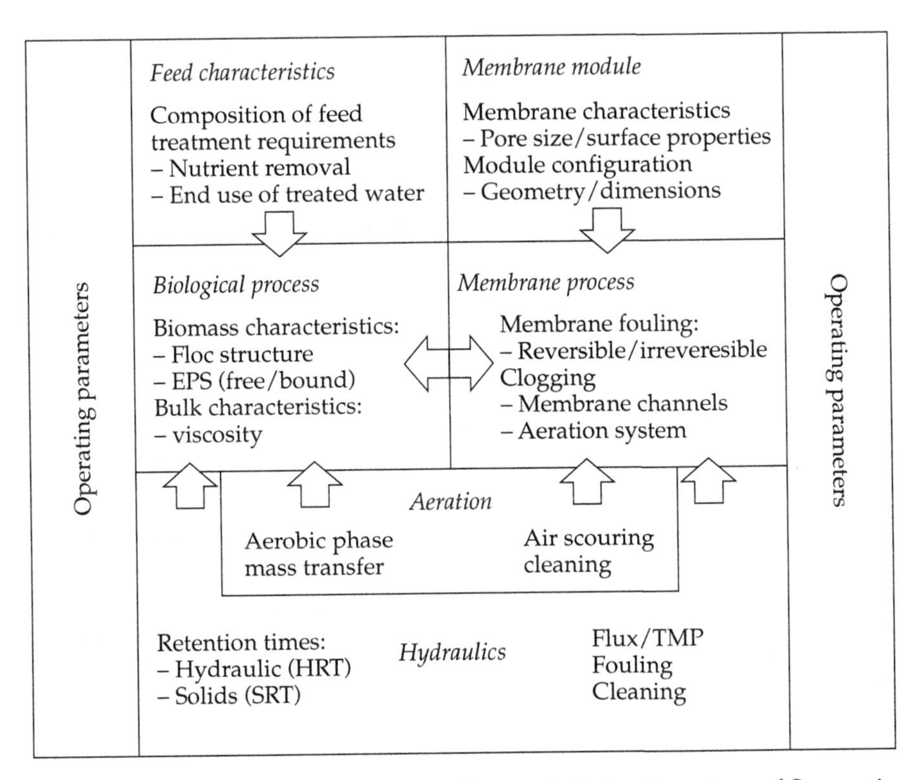

FIGURE 7.1 Example of hollow-fiber membrane sludging (courtesy of Separation Processes, Inc.).

FIGURE 7.2 Example of flat-sheet membrane sludging (Mason et al., 2010).

pretreatment processes (screens, in particular) or improper operation of the membrane air scour system.

- Chemically reversible—foulants strongly attached to the membrane surface and in the membrane pores; these are typically removed using chemicals.
- Irreversible—foulants that cannot be removed from the surface or pores of the membranes; these foulants contribute directly to the gradual decline in membrane flux.

For a more detailed discussion of foulants and fouling mechanisms, refer to Chapter 2. Table 7.2 summarizes both physically reversible and chemically reversible fouling types and common cleaning strategies.

Manufacturers often recommend a maximum lifetime of chemical dose exposure (i.e., mg/L-hr) for membranes, beyond which membrane performance can

TABLE 7.2 Summary of membrane fouling and cleaning strategies.

Type of fouling	Type of contaminant	MBR contaminant components	Type of cleaning chemical used	Typical chemicals
Physically reversible		Stringy material	None—physically removed by hand	N/A
		Materials that bypass preliminary treatment		
Chemically reversible	Organic	Biogrowth	Oxidant	Sodium hypochlorite*
		Smaller substances excreted by microorganisms		
		Dissolved organic particles		
	Inorganic	Precipitation (metals, calcium, etc.)	Acid	Citric acid*
		Oxides formed in presence of common coagulants		Oxalic acid
				Hydrochloric acid

*Denotes most common chemical used.

deteriorate, potentially reducing the life expectancy of the membrane. Alternative cleaning chemicals and procedures continue to be evaluated and developed (i.e., hydrogen peroxide for organic contaminants); it is highly recommended to continuously reexamine cleaning chemicals and procedures used with the respective membrane manufacturer.

Typically, chemical cleans are performed without removing the membranes from the membrane basin and mixed liquor and require removing the basin from service; therefore, membrane cleaning activities should be taken into account when evaluating the redundancy required in membrane train design and layout. Occasionally a separate, off-line tank that is sized to accommodate one or two large membrane subunits is provided. This allows for the ability to remove a large membrane subunit from the process train for cleaning; it also allows operations staff to evaluate alternative cleaning chemicals and/or strategies.

In general, there are two types of chemical cleaning: maintenance and recovery. Both of these types are more aggressive than the cleaning provided by the air scour and routine backwash and relaxation cycle. Initially, the frequency and duration of these cleanings should be performed as recommended by the manufacturer; however, the overall cleaning strategy should be adjusted with operational experience based on particular wastewater characteristics, bioreactor operation and performance, membrane loading and performance, and so on. It should be noted that some manufacturers do not require maintenance cleans while others do not require recovery cleans, although this is typically only a difference in terminology. Table 7.3 summarizes typical cleaning strategies for several types of MBR systems. It should also be noted that terms related to membrane cleaning can vary between manufacturers as shown in Table 7.4.

2.1.1 *Maintenance Cleaning*

Maintenance cleans are intended to enhance, through a more rigorous effort, the cleaning performed via air scour and backwashes and help decrease the frequency of longer recovery cleans. Typically, the targeted foulant is organic, and an oxidant, such as sodium hypochlorite, is used. In some instances, it may also be necessary to perform a maintenance clean to remove an inorganic foulant, effectively doubling the time to complete the maintenance clean. Often, inorganic fouling progresses at a lower rate than organic fouling, allowing for a lower cleaning frequency rate for inorganics.

Maintenance cleaning procedures vary by manufacturer. Tables 7.5 and 7.6 show the representative procedures for submerged hollow-fiber and submerged flat-sheet

Table 7.3 Summary of cleaning strategies for several representative MBR manufacturers.

Manufacturer	Membrane configuration	Cleaning strategy		
		Type	Frequency	Duration
Hollow fiber no. 1	Submerged	Maintenance	1–7 days	30–60 minutes/chemical
		Recovery	4–6 months	4–8 hours/chemical
Hollow fiber no. 2	Submerged	Maintenance	7 days	60 minutes/chemical
		Recovery	3 months	6–8 hours/chemical
Hollow fiber no. 3	Submerged	Maintenance	7 days	30–120 minutes/chemical
		Recovery	6–12 months	4–8 hours/chemical
Flat sheet no. 1	Submerged	Maintenance	3 months	2–4 hours/chemical
		Recovery	N/A	N/A
Flat sheet no. 2	Submerged	Maintenance	N/A	N/A
		Recovery	3–6 months	2 hours/chemical
Hollow tube no. 1	External	Maintenance	N/A	N/A
		Recovery	2 months	60–90 minutes/chemical

Table 7.4 Cleaning terminology cross-references.

Cleaning Event	Alternate Terminology
Backwash	Backpulse
Maintenance clean	Enhanced backwash
	Clean-in-place Regeneration clean
Recovery clean	Intensive cleaning
	Clean-in-place
Mechanical clean	Manual cleaning
	Hand cleaning
	Physical cleaning

Table 7.5 Maintenance clean—hollow-fiber MBR system.*

1. Check the cleaning chemical system, ensure adequate amounts of chemical are available and diluted to the correct concentration.
2. Isolate the appropriate membrane basin and stop permeation of the membranes.
3. Backwash through the membranes with chemical for 1–5 minutes.
4. Relax the membranes for 4–15 minutes.
5. Repeat Steps 3 and 4 (the number of times will vary depending on the manufacturer, typically 3–10 times).
6. Backwash with permeate (i.e., no chemical) to flush the chemical feed line.
7. Run air scour for 4–20 minutes to neutralize the chemical solution in the membrane tank. Some systems may also require that the permeate pump be on, returning the permeate to the membrane tank.
8. Continue air scour and allow mixed liquor to enter and exit the membrane tank for 1–30 minutes to ensure the chemical is neutralized (this step is not always required and varies with the membrane manufacturer).
9. Restart permeation/return membrane tank to service.
10. Repeat steps 8 and 9 (the number of times will vary depending on the manufacturer).

*The membrane tank is not drained during this procedure.

Table 7.6 Maintenance or recovery clean—flat-sheet MBR system.*

1. Check the cleaning chemical system, ensure adequate amounts of chemical are available and diluted to the correct concentration.
2. Isolate the appropriate membrane basin and stop permeation of the membranes.
3. Confirm permeate valve is closed.
4. Continue the air scour.
5. Inject the cleaning solution into the permeate line; the solution will be backflushed through the membrane.
6. After injecting the specified amount of chemicals, stop chemical injection.
7. Allow the membranes to soak for 1–3 hours.
8. Adjust air scour, if necessary, to control potential foaming.
9. Open the permeate valve and return the membrane train to service.
10. Return the permeate (with the spent cleaning solution) to a point upstream in the treatment plant process (i.e., headworks).
11. The cleaning solution should be flushed after a couple of cycles and the permeate allowed to once again be discharged.

*The membrane tank is not drained during this procedure.

membrane systems, respectively. The procedures in these tables represent maintenance cleans using sodium hypochlorite; although the procedures are similar when using an acid as the cleaning chemical, the operator should be careful to note the differences. A manufacturer's recommendations for cleaning procedures should always be followed; the manufacturer should be consulted prior to adopting any significant changes. It is also important to note that flat-sheet membrane systems require less frequent cleaning than other types of membrane configurations, typically only either a maintenance or recovery clean, but rarely both.

2.1.2 *Recovery Cleaning*

Occasionally, a more aggressive cleaning is required to recover membrane permeability. Recovery cleans are typically triggered when TMP increases or the permeability decreases by a certain amount. Manufacturers may recommend performing a recovery clean on a set frequency even in the absence of TMP and/or permeability triggers.

Similar to maintenance cleans, recovery cleaning procedures vary by manufacturer; however, the differences can be more significant while the goal remains the same. Table 7.7 shows the representative procedure for submerged hollow-fiber membrane systems. Flat-sheet membranes complete a high-strength chemical soak cleaning that is similar to the objective of a recovery clean in hollow-fiber systems. This is variously referred to as either a *maintenance clean* or a *recovery clean* depending on the flat-sheet membrane supplier; therefore, refer to Table 7.6 for the representative recovery clean procedure for flat-sheet membrane systems.

2.1.3 *Physical Cleaning*

Occasionally, small and large membrane subunits may require physical cleaning. This is typically caused by the buildup of solids in the small membrane subunits. Solids can build up over time between the membrane fibers or flat sheets, concentrating near the top of the subunit where the fibers enter the top support of the subunit or between the flat sheets. Referred to as *membrane sludging*, the membranes can be desludged by gently washing the solids from the membrane surfaces with water and/or by manual removal. The solids may include both compacted activated sludge and stringy material such as hair; maintenance staff should also be vigilant for hazardous materials such as hypodermic needles. Manual removal can be performed with large membrane subunits in place or after removing the large membrane subunits from the membrane basin. Manual removal with large membrane subunits in

TABLE 7.7 Recovery clean—hollow-fiber MBR system.

1. Check the cleaning chemical system; ensure adequate amounts of chemical are available and diluted to the correct concentration.
2. Isolate the appropriate membrane basin and stop permeation of the membranes.
3. Continue air scour for 5–10 minutes.
4. Drain the membrane tank.
5. Backwash without chemical to partially fill the membrane tank.
6. Air scour for 5–10 minutes.
7. Drain membrane tank and hose it down if possible.
8. Backwash with chemical for 1–2 minutes.
9. Allow the membranes to relax 4–5 minutes.
10. Repeat steps 8 and 9 until the membrane tank is filled to the half-way point (approximately 20–30 times).
11. Backwash without chemical to flush chemical solution from the chemical feed lines and to fill the membrane tank up to the cleaning level (approximately 10–15 minutes).
12. Allow the membranes to soak for 4–6 hours with intermittent aeration (15–30 seconds every 2 hours). This soaking step can often be extended by hours.
13. Run the air scour for 4–6 minutes to neutralize the chemical solution in the membrane tank.
14. Continue air scour and allow mixed liquor to enter and exit the membrane tank for 1–30 minutes to ensure the chemical is neutralized.
15. Restart permeation and return membrane tank to service.

place requires entering the membrane basin (i.e., confined space entry); therefore, it is important that proper safety measures are followed. Caution should also be taken during manual removal after removing the large subunits from the membrane basin as accumulated solids can increase membrane subunit weight dramatically. A short chemical soak following a physical cleaning can also be beneficial to the membranes.

2.1.4 *Membrane Air-Diffuser Cleaning*

Air diffusers located below large membrane subunits occasionally foul, similar to air diffusers located in the bioreactor. Fouled diffusers can lead to uneven and reduced air scouring and may result in increased backpressure at the membrane blowers. Fouled diffusers can be identified by observing the air pattern in the membrane basin, monitoring membrane flux and permeability and any changes in membrane cleaning frequency, or by draining the membrane basin and directly inspecting the diffusers.

Cleaning the diffusers typically requires draining the membrane basin, removing the large membrane subunit(s) from the membrane basin, or using a manual water flush. The two common types of buildup, organic (biological) and inorganic scaling, can be removed using either chemicals and/or physical cleaning (the procedure varies depending on the membrane manufacturer's recommendations). If fouling of the diffusers is evident, blower performance should also be reviewed.

2.2 Identifying and Replacing Damaged Membranes

Small membrane subunits may periodically become damaged, typically from the unintended introduction of foreign materials into the membrane basins either as a result of construction or failure of upstream processes to adequately protect the membranes. The damage may affect the membrane material directly, such as a leaking fiber (hollow-fiber-type systems) or an incision in the flat sheet (flat-sheet-type systems) or it can affect the hardware components of the small or large membrane subunits themselves (i.e., a leaking o-ring, permeate hose and tubing). The damage may not affect membrane performance immediately and can be addressed during the next visual inspection, which is typically scheduled annually (or quarterly during the first year); however, when damage affects performance, quickly identifying and repairing or replacing the damaged membrane in a timely fashion is critical to maintaining desired system performance levels. Several methods are available to identify and repair or replace damaged membranes and can vary depending on the type of membrane configuration and manufacturer; therefore, it is important to consult the manufacturer's O&M manual. These methods are as follows:

- Identifying the damaged membrane subunit—often, the first indication of membrane damage is an increase in permeate turbidity above 0.2 nephelometric turbidity units. The damaged large membrane subunit can be identified by measuring the permeate turbidity from each membrane tank and then by stopping permeation one by one from each large membrane subunit and observing the effect on overall permeate turbidity. It is important to note that because of the high solids content in membrane tanks, hollow-fiber-system membranes frequently correct themselves; the solids in MLSS will plug the damaged fiber because of its small inner diameter.
- Identifying the cause of damage to the large membrane subunit—visual inspection of the large membrane subunit may identify a leaking hose, pipe, o-ring, and so on. Further investigation can involve performing a clean water

test, performing a bubble-point test (hollow-fiber membranes only), removing the large membrane subunit from the membrane tank, and performing a more detailed visual inspection. These activities can be performed using a dip tank to avoid removing an entire membrane train from service.

- Repairing or replacing the membrane—depending on the type and extent of damage, the small membrane subunit can be repaired per manufacturer's instructions. Fibers can be plugged, pinned, tied, or removed (hollow-fiber membrane), and the flat sheet can be replaced (flat-sheet membrane). Replacement of an entire large membrane subunit is unusual and unlikely.

2.3 Instrument Calibration

Membrane bioreactor systems are highly automated and proper operation relies on both online instrumentation as well as periodic (daily, weekly) grab sample analysis. Therefore, it is critical to ensure that data collected by these methods is reliable and accurate. Initially, calibration activities should follow the schedule recommended by the manufacturer; however, the schedule will likely be modified as O&M staff gain experience with their particular system. Instrumentation may include online transmitters, gauges, or laboratory equipment to monitor pressure, temperature, dissolved oxygen, flow, level, pH, COD, solids concentrations, ammonia, phosphorus, turbidity, and switches that monitor air flow, level, and so on. It should also be noted that the calibration schedule should also verify the accuracy and proper operation of the data collection system (i.e., supervisory control and data acquisition [SCADA]), chemical metering and feed systems, valve operation and position indicators, and so on. It cannot be emphasized enough how important properly calibrated and operating instrumentation is to the success of an MBR system in meeting expected performance criteria.

2.4 Membrane Storage, Handling, and Replacement

The most important thing to remember when storing and handling membranes is to always follow the manufacturer's requirements; failure to do so can reduce the effectiveness of the membrane or cause irreparable damage and, at the very least, void the warranty. Although requirements vary by manufacturer for the storage and handling of the membranes, a lot of similarities exist . These include the following:

- As a general rule, membranes should never be allowed to dry out. New membranes typically are packaged with a preservation fluid that helps extend their shelf life. Although the membranes can be allowed to stand in the dry for a short period (typically less than a few hours), taking a membrane basin out of service for an extended period typically requires refilling the basin with clean water and adding a disinfectant. If a basin is to be removed from service for a long period, the manufacturer should be consulted; however, membranes can be repackaged with preservation fluid. (Although disposal of preservation fluid could be problematic in the past because of its hazardous nature, preservation fluids with biodegradable properties such as glycerin have been developed.) Proper safety and care should be taken when handling these chemicals.

- The air scour requires the membranes to have a certain level of robustness; however, they can be damaged by sharp or heavy objects. Therefore, it is important, to be careful when working around them during construction and maintenance activities.
- Membranes should never be allowed to freeze.
- During operation, efforts are ongoing to minimize the fouling of membrane pores by constituents in the mixed liquor (organic or inorganic). Although fouling is not as great of a concern when membranes are out of service, it is still important to be aware of the potential (e.g., oily hands coming in contact with the membrane surface).

Membrane replacement may be required periodically; this may involve replacement of an individual small membrane subunit or the replacement of a large membrane subunit. Replacement of a small membrane subunit is occasionally required when a membrane has been damaged and cannot be repaired. Generally, the large membrane subunit is removed from the membrane basin to allow replacement of one of its small membrane subunits. The procedure is similar when adding additional small membrane subunits to a large membrane subunit that is not fully "loaded." Replacement of a large membrane subunit can involve two significant components: the small membrane subunits and the frame that contains the small membrane subunits. Replacement of the frame is rare; replacement of all the small membrane subunits is more typical and generally occurs at the end of the membrane's expected life. The "end of a membrane's expected life" is loosely defined as when, because of irreversible fouling, the membrane is unable to perform its duties within certain parameters (i.e., clean flux, cleaning frequency). It is possible that a contaminant (i.e., oil, grease, polymers) in the wastewater fouls the membranes, prematurely ending the membrane's life.

There are certain "wear" parts for some membranes that should routinely be inspected and may require periodic replacement. Some of these items include hose clamps, permeate tubing, and mounting hardware.

2.5 Ancillary Systems

Membranes are the heart of an MBR system and are supported by several ancillary equipment systems whose proper operation directly affects membrane and MBR performance. An MBR supplier typically does not manufacture all of the supplied equipment, but provides a "system package." As such, it is important to consult both the MBR supplier's O&M manual and individual O&M manuals supplied for equipment items (i.e., pumps, blowers). Improper maintenance of these ancillary systems can adversely affect operation of the MBR system and potentially shorten membrane life. It is also important to properly maintain equipment for systems upstream of the MBR because all of those systems help to protect the membranes in some manner, whether directly (e.g., screens, grit systems, proper operation of the activated sludge system) or indirectly (e.g., grit classifiers, electrical feed system).

2.5.1 *Membrane Basins*

Membrane basins for immersed systems that house large membrane subunits should be periodically removed from service, drained, and cleaned for inspection. The basins should be inspected for mechanical and structural damage and a structural engineer consulted for any necessary repairs. Because some of the membrane cleaning procedures involve the use of sodium hypochlorite and various acids, a protective coating is typically applied to the concrete. This coating should be inspected and repaired as necessary.

2.5.2 *Rotating Equipment*

Routine maintenance schedules should be followed for membrane blowers, membrane pumps, and other equipment (i.e., membrane basin drain pumps, chemical transfer and feed pumps). Although each system has an important role in the MBR system, none is as critical as membrane blowers. For immersed systems with membrane air scour, operating membranes without air for any length of time can lead to irreversible fouling and potentially shorten membrane life.

2.5.3 *Compressed Air System and Cyclic Valves*

Several immersed systems have air scour strategies that provide air to the membrane basins intermittently. The intermittent operation reduces the overall air demand of the membranes and subsequently the power demand and cost to operate the blowers. The cost savings can be substantial; however, the intermittent operation relies on valves in the air scour line, which can cycle open and closed several times every minute. Typically, these valves are air-operated and require rebuilding every so often. Additionally, leaks in the compressed air system piping should be addressed promptly because leaks can potentially accumulate to the point where the load on the air compressors increases beyond their capacity. Routine maintenance of the compressed air and air-dryer systems can reduce nuisance failures of the cyclic valves and extend their life.

2.6 Manufacturer Support

Support from a manufacturer can vary depending on the services obtained by the owner of the MBR system. However, it is important for the manufacturer to be involved during delivery and installation of equipment, particularly during startup of the new system. A service and support contract is commonly executed between the manufacturer and the owner for the first year or two of operation, especially for new users of MBR systems. With the advent of SCADA systems, manufacturers typically can monitor an operation from a remote location. Additional manufacturer support beyond the first few years is also an option and varies according to the comfort level of the owner.

3.0 OPTIMIZATION

3.1 Process

As discussed in previous chapters, one of the advantages of an MBR system is that it allows high concentrations of MLSS and increased SRT, which enable high BOD_5 removal efficiency and nitrification compared to CAS systems. To date, the high cost of membranes and membrane fouling have deterred many people in the industry from MBRs. In the past few years, considerable research has been conducted to understand membrane fouling in more detail. Much of this research has centered on optimizing MBR operating parameters, such as SRT, HRT, dissolved oxygen, and F/M, to alter activated sludge characteristics and control various membrane fouling types. Specific biological control strategies are discussed in more detail in previous chapters of this MOP. Nevertheless, optimization of sludge characteristics for membrane fouling control will continue to be an important topic for future research efforts and application of MBRs (Meng et al., 2009).

3.2 Chemical Usage

Chemicals are used throughout the treatment process for various purposes including coagulation and precipitation, carbon source for the activated sludge process, disinfection, membrane cleaning, and membrane flux enhancement. The quantity of chemical used directly translates to operating costs; therefore, minimizing chemical usage and dosage rates will minimize costs. It is beyond the scope of this section to adequately describe the many strategies available for minimizing chemical usage throughout the treatment plant. Water Environment Federation's *Operation of Municipal Wastewater Treatment Plants* (WEF, 2007) and other sources can provide more information on this topic. In terms of the membranes themselves, the largest chemical demand is related to membrane cleaning activities. There is a tradeoff between minimizing the type and frequency of cleanings and the long-term longevity of the membranes; however, the most optimal use of chemicals can be achieved through operator experience with their system and by consulting with membrane manufacturers.

3.3 Operational Strategies for Energy Management

A key area of focus within the MBR industry is energy reduction and optimization. Currently, the energy requirement for an MBR typically exceeds that of a CAS plant. As shown in Figure 4.6 (Chapter 4), more than 75% of the energy consumed in an MBR system is from membrane and biological aeration systems. As a result, it is important to consider energy optimization during design. Chapter 4 discusses several design elements for optimizing energy usage. Operational elements to consider include the following:

- Operate the air scour rate based on the instantaneous flux to allow lower air flow at lower flux. This strategy can be an effective method of flow-pacing the air scour rate with permeate flow for some systems;
- Maximize, if possible, the return of oxygen-rich mixed liquor to aerobic zones in the biological process (see Chapter 4, Sections 4.6 and 5.6, for further discussion);
- Operate with membrane performance enhancers (metal salts or polymer) to reduce the overall membrane footprint and, therefore, the required air scour;
- Operate within the recommended range of SRTs for MBRs to minimize endogenous respiration, associated aeration demand, and membrane fouling;

- Optimize the air scour strategy working closely with the membrane manufacturer (this may not be an option with every MBR);
- Optimize the membrane cleaning strategy working closely with the membrane manufacturer;
- Maintain operation with the minimum number of membranes and membrane trains required online. The other membranes and membrane trains are in standby mode, but are cycled through using an alternation sequence. This will reduce the overall air scour requirement (membranes in standby mode typically require some air scour, although at a reduced rate), but could increase the cleaning frequency and complexity of operations; and
- Maximize use of online instrumentation for a greater level of control of the activated sludge and membrane processes.

It should be noted that not all of the aforementioned energy-saving strategies are feasible with all MBR systems.

3.4 Innovative Approaches

As the MBR industry continues to grow, new and innovative approaches are being investigated and developed by manufacturers and owners to reduce capital, operating, and energy costs. Prior to implementation, these strategies should be carefully evaluated, weighing each strategy's track record against the potential benefits and risks with the facility's operational history. These approaches include the following:

- Carefully considering operational flexibility in the system design to minimize the effort in bringing membrane basins and/or other equipment (i.e. membrane blowers, permeate pumps, etc.) on and offline.
- Adjusting or flow-pacing the membrane air scour rates to coincide with the wastewater plant flowrate. Although most systems incorporate this concept to some degree, work continues in this area in an effort to further optimize the process.
- "Recycling" membranes; examples of this approach include the Anthem Water Campus (Maricopa County, Arizona), which installs "used" membranes from the water treatment plant in the wastewater and MBR plant, and the Cauley Creek Water Reclamation Facility (Fulton County, Georgia), which installs used MBR membranes in the sludge-thickening system.

- Using a flux-enhancement chemical or polymer to handle peak flows greater than expected and short-term use of chemicals during peak flow events; capital and operating costs of chemicals should be compared to potential infrastructure savings (i.e., additional flow equalization and/or reduction in membrane area).

4.0 TROUBLESHOOTING

Table 7.8 presents a troubleshooting guide that can be used as a quick reference. For detailed troubleshooting of MBR systems, consult the MBR manufacturer's O&M manual. If any of the problems discussed in this section or other difficulties persist,

TABLE 7.8 Troubleshooting guide.

Condition	Possible cause	Potential remedies
Permeate		
Quality		
The permeate turbidity or TSS is above the specified setpoint (i.e., permeate is "cloudy")	Faulty operation of instrumentation	Confirm proper operation of control systems and instrumentation, clean, replace, recalibrate.
	Small membrane subunit—membrane fibers or sheets are damaged	Repair or replace the membrane fibers or sheets.
	Large membrane subunit—damage to the membrane assembly	Identify damaged part and replace (i.e., leaking tube, pipe, o-ring, gasket).
	Permeate piping leak	Identify the leak and repair as required.
Quantity		
Lower permeate flow than expected	Mechanical equipment failure	Ensure proper operation of permeate pumps (pumped system) or the permeate control valve (gravity system).

(continued)

Table 7.8 Troubleshooting guide (*continued*).

Condition	Possible cause	Potential remedies
	Incorrect operation of permeate piping and/or valving	Check permeate piping for leaks; confirm permeate valves are in their correct position.
High TMP and lower permeability than expected	Membrane sludging	Remove materials from small and large membrane subunits and confirm proper operation of pretreatment systems.
	Mixed liquor characteristics are contributing to poor filterability	Test the filterability of the mixed liquor; review and improve operating conditions of the activated sludge system as necessary.
	Desired operating flux or flowrate is higher than design	Adjust operational flux or flowrate to match the design flux or flowrate.
	Membrane surface is fouled	Review recent cleaning records, review MBR manufacturer's O&M manual, and initiate appropriate cleaning procedure.
	MLSS is high SRT too low (below that needed for nitrification) Insufficient dissolved oxygen in aerobic zones	Improve the characteristics of sludge by adjusting the activated sludge process operation.
Uneven permeate and/or MLSS flow between membrane basins	More recent completion of maintenance or recovery cleanings for some of the membrane basins	Review membrane cleaning records.
	Uneven flow distribution to each membrane basin	Confirm proper operation of flow distribution system (slide gates, weirs, pumps/control valves, etc.).
Membrane air system		
Uneven aeration of large membrane subunit	Membrane diffuser clogging	Wash or clean the diffuser(s).

(*continued*)

TABLE 7.8 Troubleshooting guide (*continued*).

Condition	Possible cause	Potential remedies
Uneven aeration in membrane train	Membrane diffusers or large membrane subunits are not level	Level the equipment.
Air scour rate fails to reach the specified values	Improper operation of membrane blower(s) and/or air scour system, malfunctioning diffuser(s)	Repair or replace the blower(s); review and repair air piping as necessary; repair the diffuser(s).
Air scour is not properly cycling	Improper operation by air cycling valves	Repair or rebuild air cycling valves; review operation of the compressed air system.
Membrane cleaning		
Increased frequency of cleaning	Mixed liquor characteristics are contributing to poor filterability	Test the filterability of the mixed liquor; review and improve operating conditions of the activated sludge system as necessary.
	Improper operation of membrane cleaning system	Confirm cleaning system is operating correctly; controls, cleaning cycles or durations, chemical feed, and dosing systems.
	"Over-permeation" of membranes	Review MBR flow records.
Cleaning fails to recover flux, permeability of membranes	Improper operation of membrane cleaning system	Confirm cleaning system is operating correctly; controls, cleaning cycles, durations, chemical feed, and dosing systems.
	Improper operation of permeate pumps, piping, and control valving	Repair equipment as necessary.
	Irrecoverable fouling of membranes	Review operating history with the goal of identifying contributing factor(s) to fouling; perform membrane autopsy on small membrane subunit.

(*continued*)

TABLE 7.8 Troubleshooting guide (*continued*).

Condition	Possible cause	Potential remedies
Activated sludge system		
Effluent quality does not meet specified values	Mixed liquor characteristics are not within proper operating standards	Improve the operating conditions of the activated sludge system.
Mixed liquor concentration is too high or too low	Sludge wasting rates are incorrect	Review operation of the activated sludge process and revise the sludge wasting rates as necessary.
Foaming		
Excessive foaming in activated sludge and/or membrane basins	Mixed liquor characteristics are not within proper operating standards	Improve the operating conditions of the activated sludge system.
	Membrane air scour system is overaerating	Review operation of the membrane air scour system.
	Foam control strategies are not operating correctly	Review foam removal equipment, anti-foam chemical dosing system.
	MBR and activated sludge system are undergoing startup	Control and remove foam as necessary until startup is completed and system reaches a steady state.

the operator should contact the MBR equipment manufacturer for additional assistance. It is important to consult all necessary safety information prior to troubleshooting and ensure that only qualified personnel are troubleshooting MBR systems. Mechanical or electrical maintenance should only be performed by qualified personnel. In the event of any differences between this troubleshooting guide and the MBR manufacturer's O&M manual, the MBR manufacturer's information should always take precedence.

5.0 LESSONS LEARNED

Although use of MBR plants has grown rapidly worldwide, operational issues associated with MBRs are not fully understood by operators or design engineers. To determine the challenges faced during day-to-day operations, an informal survey of approximately 30 operating MBR facilities in North America was conducted. The plants surveyed included both hollow-fiber and flat-sheet membranes and plants with flow capacities ranging from 1500 m^3/d to 38 000 m^3/d (0.4 to 10 mgd). This section summarizes the most common suggestions or lessons learned from the survey, which each have the goal of ensuring a more operations-friendly facility. Many of the operations staff contacted during the survey also had experience in operating a conventional treatment plant.

5.1 Pretreatment

During the survey, many operators stated that proper influent screening (i.e., the coarse and fine screens) is the most important part of an MBR treatment plant. They also stated that foreign debris (hair, in particular, and generally with hollow-fiber membranes) in membranes can cause the greatest operating headaches when the fine screens are not operating correctly or, in some instances, not included in the process treatment train. Removal of foreign debris from membranes is typically accomplished by hand, requiring either removal of the large membrane subunit from the membrane tank or draining the membrane tank completely.

5.2 Membrane Tanks

Membrane tanks should be configured and constructed with the following factors or considerations in mind:

- *Foaming*—a higher incidence of foaming is common in MBRs; therefore, sufficient freeboard in the membrane and activated sludge tanks is critical. Design of the system should allow some method for handling the foam (i.e., spray system, removal). Foam removal mechanisms typically need to be sized for more than would be encountered in a CAS system.
- *Membrane air scour*—careful consideration of membrane air scour control is important (it is important to "keep it simple"). Attempts to control the air scour using the dissolved oxygen level have been problematic.
- *Cleaning*—the survey revealed that many MBR plants do not have an easy way for operators to drain the membrane tanks for cleaning. Operators suggested that all MBR designs should have a sump in the membrane tanks and some type of pumping system to provide quick and easy access for drainage.

- *Membrane removal*—a bridge crane for removal of membranes was deemed to be important. It is important to note that the bridge crane must be large enough to handle a wet, fully submerged membrane module.
- *Offline test/cleaning tank*—the provisions of an offline tank, sized to accommodate one or two large membrane subunits, was desirable, particularly for large-capacity plants.
- *Tank covers* (see Figure 7.3)—an operator should consider covering membrane tanks with solid covers to eliminate possible introduction of debris, thereby bypassing the protection of upstream pretreatment processes. The potential of unwanted objects entering the membrane tanks can vary based on the site location and surrounding area. One MBR plant surveyed has an occasional issue with unwanted tumbleweeds finding their way into the membrane tanks (removal requires draining the tank). Venting of the airspace underneath the covers is necessary and needs to be considered.
- *Tank coating systems*—proper selection and application of the concrete coating protection system is critical during construction otherwise coating may flake off and damage the membranes.

FIGURE 7.3 Membrane tank covers and overhead crane at Broad Run Water Reclamation Facility, Loudoun County, Virginia (courtesy of Hazen & Sawyer, PC).

- *Membrane installation*—careful installation and startup of the small and large membrane subunits is important to proper operation. It is also important to ensure that all of the manufacturer's installation and check-out requirements are addressed.
- *Tank influent baffles*—consider installing tank influent baffles for energy dissipation to protect the membranes from high velocities.

5.3 Training

Proper training is critical because membranes remain a relatively new technology for most operators; indeed, it is rare for an operator to have experience operating an MBR. The membrane is viewed as a "black box" technology and most survey respondents indicated that they were initially intimidated by the system, much more so when compared to other "first experiences" during their career. Survey respondents had similar responses when discussing a significant point they try to convey when training new staff in MBR operation. That is, that although MBRs are different from anything the trainee has likely operated in the past, they are no more difficult to operate once trained.

5.4 Membrane Bioreactor Strengths

Operators surveyed stated that the best feature of the MBR is its consistent effluent quality. With an MBR they were much less concerned with meeting the effluent limits as when they were operating a conventional treatment plant. They also stated that because most MBR systems are highly automated, they require fewer full-time staff onsite. Therefore, the MBR process was, in general, easier to operate than a conventional plant.

5.5 Membrane Bioreactor Weaknesses

Several operators surveyed stated that a crucial weakness of an MBR system is power consumption, which is primarily caused by the membrane air scour, the higher RAS flowrate, and, to some extent, the lower oxygen transfer efficiency in the activated sludge caused by the high mixed liquor concentration. Therefore, the operators suggested that a good strategy to reduce air scour be implemented during the design stage.

6.0 OPERATIONAL CONSIDERATIONS THAT DIFFER FROM CONVENTIONAL ACTIVATED SLUDGE SYSTEMS

Knowledge and experience gained through operation of a CAS system is valuable for first-time MBR plant operators. However, there are differences between the two types of systems. The following list highlights important points for operators making the transition to an MBR plant:

- Effluent TSS quality is consistent and generally not dependent on the MLSS concentration, sludge settling properties (i.e., SVI), and composition of the bacterial population;
- Membranes have a finite hydraulic throughput limitation (i.e., secondary clarifiers allow for a higher peak flowrate);
- Extended high-flow events can result in additional membrane maintenance (i.e., cleanings) and may potentially irrecoverably foul the membranes (reducing filtration capacity);
- Membrane bioreactor systems are typically more prone to foaming because of the smaller bioreactor surface area, higher MLSS concentration, and higher process air flowrate;
- Membrane maintenance (i.e., fouling management, membrane cleaning) is directly affected by the biological process and hydraulic relationships (some of these interdependencies are shown in Figure 7.4);
- Higher RAS rates;
- Membrane performance can be sensitive to improperly operating pretreatment systems (i.e., fine screens, grit);
- Membrane bioreactor systems typically have a higher degree of automation;
- Although the MBR process is well-developed, it has a shorter operating track record and continues to evolve; therefore, it is important to periodically compare notes with the MBR manufacturer and operations and engineering staff from other MBR treatment plants;
- Unlike clarifiers, membrane trains do not always produce permeate because of cleaning, backpulse, and relaxation events. For example, a 10-minute cycle

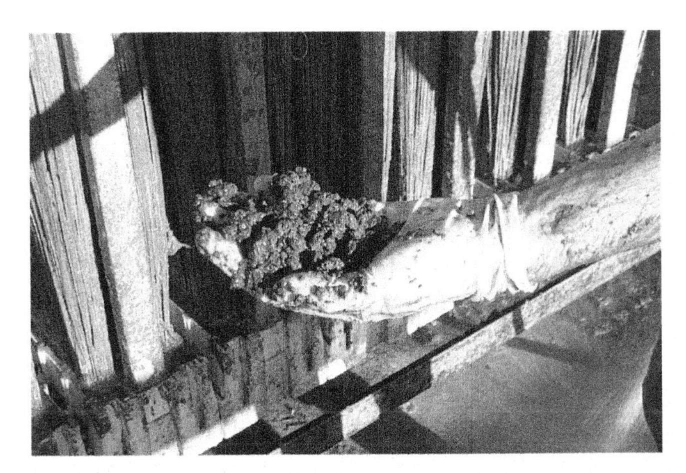

FIGURE 7.4 Key elements and interactions of the MBR process (Leiknes et al., 2009).

(9.5 minutes production plus 0.5 minutes relaxation) means each membrane train is offline (i.e., not producing permeate) for 1 hour and 12 minutes during a 24-hour period (i.e., 5% of the time).

7.0 REFERENCES

Leiknes, T.; Phataranawik, J.; Ivanovic, I. (2009) Prospects and Potentials of Biofilm-MBRs for Municipal Wastewater Treatment. *Proceedings of the 82nd Annual Water Environment Federation Technical Exhibition and Conference* [CD-ROM]; Orlando, Florida, Oct 10–14; Water Environment Federation: Alexandria, Virginia.

Meng, F.; Chae, S. R.; Drews, A.; Kraume, M.; Shin, H. S.; Yang. F. (2009) Recent Advances in Membrane Bioreactors (MBRs): Membrane Fouling and Membrane Materials. *Water Res.*, **43**, 1489.

Water Environment Federation (2006) *Membrane Systems for Wastewater Treatment;* Water Environment Federation: Alexandria, Virginia.

Water Environment Federation (2007) *Operation of Municipal Wastewater Treatment Plants*, 6th ed.; Manual of Practice No. 11; Water Environment Federation: Alexandria, Virginia.

8.0 SUGGESTED READINGS

American Water Works Association (2005) *Microfiltration and Ultrafiltration Membranes for Drinking Water, Manual of Water Supply Practices M53*, 1st ed.; American Water Works Association: Denver, Colorado.

Côté, P.; Brink, C.; Adnan, A. (2006) Pretreatment Requirements for Membrane Bioreactors. *Proceedings of the 79th Annual Water Environment Federation Technical Exhibition and Conference* [CD-ROM]; Dallas, Texas, Oct 21–25; Water Environment Federation: Alexandria, Virginia.

Enviroquip/Kubota Corporation (2009) *MBR Operation & Maintenance Manual.* Enviroquip (a division of EIMCO Water Technologies): Austin, Texas.

Kruger, Inc. (2010) *Neosep™ MBR System: Membrane Module Operation and Maintenance Manual.* Kruger, Inc.: Cary, North Carolina.

Mason, S.; Ewert, J.; Ratsey, H.; Sears, K.; Beale, J. (2010) Flat Sheet Membrane Bioreactor Operational Experiences – A New Zealand Perspective. *Proceedings of the 2010 Water New Zealand Conference & Expo;* Christchurch, New Zealand, Sept 22–24.

Merlo, R. P.; Trussell, R. S.; Hermanowicz, S. W.; Jenkins, D. (2007) Effects of Sludge Properties on the Thickening and Dewatering of Waste Activated Sludge. *Proceedings of the 2007 Residuals and Biosolids Management Conference;* Denver, Colorado, April 15–18; Water Environment Federation: Alexandria, Virginia.

National Water Research Institute (2007) *Short Course on Membrane Bioreactors;* Orlando, Florida, April 3–4; National Water Research Institute: Fountain Valley, California.

Siemens Water Technologies Corp. (2010) *MemPulse® MBR Operation & Maintenance Manual;* Siemens Water Technologies Corp.: Waukesha, Wisconsin.

Stone, M.; Livingston, D. (2008) Improvements in Cleaning Fouled MBR Membranes. *Proceedings of the 2008 IWA North American Membrane Research Conference;* Amherst, Massachusetts, Aug 10–13; International Water Association: London, U.K.

Water Environment Research Federation (2004) *WERF Project No. 00-CTS-8a: MBR Website Strategic Research*; Water Environment Federation: Alexandria, Virginia.

Water Environment Federation (2009) *Energy Conservation in Wastewater Treatment Facilities*; Manual of Practice No. 32; Water Environment Federation: Alexandria, Virginia.

Zenon Environmental Inc. (2009) *MBR Operation & Maintenance Manual*. Zenon Environmental Inc.: Oakville, Ontario, Canada.

Appendix A

Standard Membrane Bioreactor Computations

1.0 INTRODUCTION

The following section presents standard computations for membrane bioreactor (MBR) systems. The section outlines important MBR-specific design considerations and calculations, although it is not a complete step-by-step guide to MBR design. Therefore, the design example presented does not address all possible analyses, evaluations, safety factors, or design considerations. Furthermore, the design example assumes the reader has a prior understanding of the design of biological wastewater treatment systems and, therefore, does not address this aspect of MBR design. More information on the general design of activated sludge systems can be found in *Design of Municipal Wastewater Treatment Plants* (WEF et al., 2009). These standard computations focus on understanding key parameters for MBR system design. Depending on the specific parameter and the decision of the engineer and/or owner of the MBR system, these parameters may be specified by the engineer in procurement documents or provided by membrane vendors.

2.0 DESIGN EXAMPLE OVERVIEW

2.1 Design Flowrates and Maximum Monthly Loading Rates

A new MBR facility is being designed to treat influent flows and maximum monthly loading rates, as summarized in Table A.1.

TABLE A.1 Influent design flowrates and maximum monthly loading rates.

Influent flowrates		
Average annual	8 000	m^3/d
Maximum month	11 500	m^3/d
Peak day	18 000	m^3/d
Peak hour	21 000	m^3/d
Maximum month loading rates		
Biological oxygen demand (BOD)	2 500	kg/d
Total suspended solids	2 400	kg/d
Total Kjeldahl nitrogen	320	kg/d
Ammonia nitrogen	220	kg/d
Total phosphorus	45	kg/d

2.2 Design Temperature

Historical data indicate that the minimum weekly influent temperature is 10 °C; therefore, 10 °C will become the minimum design temperature for the membrane system.

2.3 Treatment Objectives

The MBR treatment system must meet effluent characteristics summarized in Table A.2 on an average monthly basis.

2.4 Overview of the Biological Reactor

The biological reactor for this application has been designed to achieve nitrification and denitrification. Because the plant must meet a total nitrogen limit of 5 mg/L, the system has been designed with two sets of anoxic and aerobic zones. The influent flow is split between the anoxic zones to supply organic carbon for denitrification. The treatment plant does not include a primary clarifier, but does include a 2-mm fine screen. The influent fats, oils, and grease (FOG) concentration is expected to be low enough to meet the vendor-specified requirement that the mixed liquor FOG concentration is less than 100 mg/L. The required reactor volume was determined to be 1960 m^3, with 25% anoxic volume. The required membrane tank volume was determined to be 500 m^3 based on the required membrane surface area and the packing density of the selected membrane product. The return activated sludge return pumping system is designed for a flowrate of 4 times the maximum monthly influent flowrate and the mixed liquor return flow returning nitrate to the first anoxic zone is designed for a flowrate of 2 times the maximum monthly influent flowrate. Figure A.1 summarizes relevant flowrates and reactor volumes. The mixed liquor concentration in the biological reactor will range from 6000 to 8000 mg/L total suspended solids

TABLE A.2 Treatment objectives.

Turbidity	≤ 0.5 NTU*
Total suspended solids	≤ 5 mg/L
Biochemical oxygen demand	≤ 5 mg/L
Total nitrogen	≤ 5 mg/L

*NTU = nephelometric turbidity units.

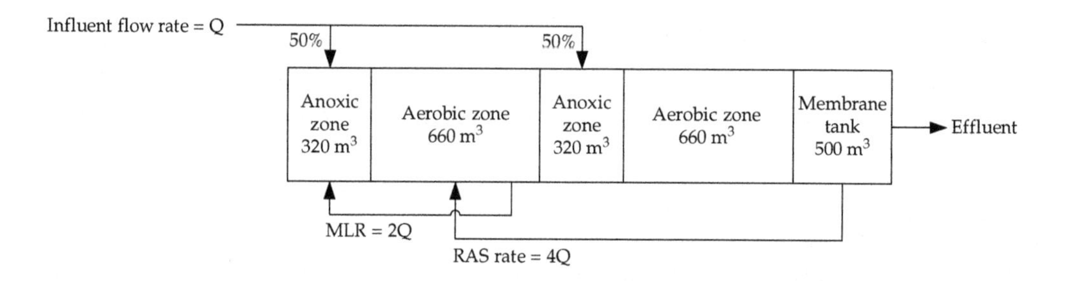

FIGURE A.1 Biological process design overview.

(TSS). The design solids retention time (SRT) is 14 days; however, the possible SRT range is from 10 to 20 days. Wastewater flows beyond the peak-day condition will be equalized in a tank upstream of the aeration basin.

2.5 Membrane Design and Redundancy Requirements

The membrane system will be designed with enough membrane area to accommodate the peak-day flow condition with one membrane train out of service; the MBR will have five membrane trains. The membrane system will also be provided with 10% spare space per membrane train to allow for the installation of additional membrane area if needed. This spare membrane area will not be installed immediately, rather, it will provide contingency for unexpected membrane fouling conditions.

3.0 STANDARD COMPUTATIONS

3.1 Membrane System Design Information

For the purpose of this design example, Table A.3 summarizes relevant membrane system assumptions. These parameters may be specified by the design engineer or provided by the membrane vendor to meet performance criteria specified by the design engineer. No attempt has been made to differentiate these items here.

3.2 Parameter Definitions

Table A.4 summarizes the parameter definitions used herein.

TABLE A.3 Assumed membrane design factors.

Membrane area per small subunit	32 m^2
Number of small membrane subunits per large membrane subunit	48
Design flux for peak-day flow based on influent flowrate	30.5 LMH*
Air scour rate for flows up to maximum monthly flowrate	10 seconds on/ 30 seconds off
Air scour rate for peak-day flowrate	10 seconds on/ 10 seconds off
Total relaxation interval	12 minutes
Relaxation time	30 seconds
Maintenance clean interval	4 days
Maintenance clean duration	60 minutes
Recovery clean interval	180 days
Recovery clean duration	8 hours
Maximum allowable solids flux to the membrane surface at the net flux of the system	325 $g/m^2 \cdot h$

*LMH = liters per meter squared per hour (standard).

TABLE A.4 Parameter definitions, abbreviations, and units.

Parameter Description	Abbreviation	Unit
Influent flowrate	Q	m^3/d
Design net flux	J	LMH
Instantaneous flux	$J_{instantaneous}$	LMH
Membrane area	A	m^2
Membrane area per small subunit	A_{SSU}	m^2
Time between relaxation	$t_{relaxation}$	minutes
Duration of relaxation	$\tau_{relaxation}$	minutes
Time between maintenance cleans	$t_{maintenance}$	minutes
Duration of maintenance clean	$\tau_{maintenance}$	minutes
Number of relaxations between maintenance cleans	N	
Online factor of the membrane system	η	

3.3 Required Membrane Area

$$J = \frac{Q}{A} \rightarrow A = \frac{Q}{J} = \frac{18\ 000 \, \mathrm{m^3/day} \cdot \frac{1000\ \mathrm{L}}{1\ \mathrm{m^3}}}{30.5 \, \mathrm{L/m^2hr} \cdot \frac{24\ \mathrm{hr}}{1\ \mathrm{day}}} = 24\ 590\ \mathrm{m^2} \tag{A.1}$$

In practice, the membrane surface area requirements for all flow conditions and operating scenarios would be assessed against the net flux at those conditions to determine the scenario that drives design of the membrane system. For the current example, the peak-flow condition drives the membrane system surface area requirements.

3.4 Required Number of Small Membrane Subunits with Ten Percent Spare

$$\text{Number of Small Subunits} = \frac{A}{A_{\mathrm{SSU}}} = \frac{24\ 590\ m^2 \cdot 1.10}{32\ m^2} = 840 \text{ Small Subunits} \tag{A.2}$$

3.5 Required Number of Large Membrane Subunits

$$\text{Number of Large Subunits} = \frac{840}{48} = 18 \text{ Large Subunits} \tag{A.3}$$

This number of large units requires 4.5 large membrane subunits per membrane tank. The membrane basin is sized to allow installation of five large membrane subunits as a provision for further expansion. Thus, the actual spare membrane area available is 22%.

3.6 Required Membrane Tank Volume

The chosen membrane product requires 20 m^3 per large subunit at the chosen packing density. With five large subunits per membrane train and five total membrane trains, the required membrane tank volume is 500 m^3.

3.7 Instantaneous, Temperature-Corrected Flux

The design flux value of an MBR system is not a measure of the actual flux of the system. Because the membrane system spends time in nonproductive modes of operation (such as relaxation), the actual instantaneous flux during membrane operation must be greater than the design value to treat the full influent flow. Table A.5 illustrates the determination of the instantaneous, temperature-corrected flux values for the peak-day flowrate. It is important to note that the pumping and piping systems for an MBR would be sized to carry the maximum expected instantaneous flowrate. It is also important to note the higher flux value for temperature correction to 20 °C. The designer should always verify the temperature that corresponds to a flux value provided by the vendor or assumed in a design calculation.

3.8 Peak-Day Solids Loading Rate

The solids loading rate at peak day must be checked to ensure that the design requirement is met. The design net flux is 30.5 LMH at peak-flow conditions. Assuming the membrane tank has a solids concentration of 10 000 mg/L TSS, the solids flux during peak flow is 305 g/m^2·h, which is less than the 325 g/m^2·h maximum provided in Table A.3. However, it is important to note that the maximum allowable solids

Table A.5 Determining instantaneous, temperature-corrected flux.

Number of relaxation cycles per maintenance clean cycle	$n = t_{\text{maintenance}} / t_{\text{relaxation}} = 5760\,\text{min} / 12\,\text{min} = 480$	(A.4)
Time spent in nonproductive relaxation mode	$= n \cdot \tau_{\text{relaxation}} = 480 \cdot 0.5\,\text{min}/cycle = 240\,\text{min}$	(A.5)
Time spent in nonproductive maintenance clean mode	$\tau_{\text{maintenance}} = 60\,\text{min}$	
Online factor	$\eta = 5460\,\text{min} / 5760\,\text{min} = 0.95$	(A.6)
Flux at 10 °C	$J = 30.5\ LMH$	
Instantaneous flux at 10 °C	$J_{\text{instantaneous},10^\circ C} = J / \eta = 32.1\ LMH$	(A.7)
Instantaneous flux at 20 °C	$J_{\text{instantaneous},20^\circ C} = J_{\text{instantaneous},10^\circ C} \cdot 1.025^{20-10} = 41.1\ LMH$	(A.8)

loading rate to the membrane would be exceeded during peak-day flows if the membrane tank solids concentration is larger than 10 600 mg/L when the peak-day influent flow arrives.

3.9 Air Scour Air Demands

The membrane supplier recommends an air scour rate of 10 Nm^3/h (i.e., normal cubic meters per hour) per small subunit under average conditions and an air scour rate of 20 Nm^3/h per small subunit under peak-flow conditions. The designer has chosen to install enough air scour for the 840 small subunits; therefore, the air scour blowers must be able to provide 8400 Nm^3/h on average and provide 16 800 Nm^3/h under peak-flow conditions. Two alternative approaches available are (1) to provide blower capacity for only the membrane area installed, not including the 22% spare, or (2) to provide blower capacity for the full five large subunits per membrane trains as a provision for future expansion, and throttle the blower, as appropriate, to deliver the required air for the installed membranes.

3.10 Membrane Permeability

Permeability is used as a measurement for determining when a cleaning cycle is needed. Permeability is measured as the temperature-corrected flux divided by the transmembrane pressure (TMP) and, therefore, simultaneously provides information about both membrane flux and TMP, which are critical operating parameters. The membrane manufacturer recommends a maximum typical operating TMP of 0.4 bar. At a temperature-corrected instantaneous flux value of 41.1 LMH at peak-day flow, the minimum allowable membrane permeability is 102.8 LMH/bar.

4.0 RECOVERY CLEANING CALCULATIONS

Performing a recovery cleaning cycle involves emptying the mixed liquor from the membrane tank and then filling it with a chemical solution for cleaning and soaking the membranes. Sizing of the various pumps used for the clean-in-place cycle is a function of the size of the membrane tank, the amount of time allocated to draining and filling the tanks, the chemical used, and the desired chemical solution strength. This section describes typical calculations used for sizing the pumping and chemical systems.

4.1 Membrane Tank Drain Pump

The membrane tank drain pumps are typically used to pump mixed liquor and spent chemical cleaning waste from the membrane tanks. For this example, the drain pumps are sized to empty the membrane tank in 30 minutes. The membrane system has five large subunits that are independently chemically cleaned. Each large subunit is housed in a tank with a volume of 100 m^3. Therefore, the drain pump has a design capacity of 3 m^3/min.

4.2 Recovery Cleaning Chemical Volume

The amount or volume of chemical needed for each recovery cleaning cycle is a function of the size of the membrane tank, the desired solution strength for the chemical cleaning cycle, and the strength of the chemical solution actually delivered or stored at the facility. The membrane manufacturer recommends twice-yearly recovery cleaning cycles, each one lasting 8 hours. The recommended chemical cleaning regime includes 1000 mg/L sodium hypochlorite and 2000 mg/L citric acid. For the purposes of this design example, the volume of sodium hypochlorite required is illustrated in Table A.6. Sodium hypochlorite is delivered to the treatment plant with a concentration 12.5%, by weight, and has a specific gravity of 1.175.

Table A.6 Determination of chemical usage for recovery cleaning.

Mass of sodium hypochlorite required per recovery clean	$= \frac{1\text{ g NaOCl}}{\text{L}} \cdot 100\text{ m}^3 \cdot \frac{1000\text{ L}}{\text{m}^3} = 100000\text{ g NaOCl}$	(A.9)
Mass of bulk chemical required per recovery clean	$= 100000\text{ g NaOCl} \cdot \frac{100\text{ g bulk chemical}}{12.5\text{ g NaOCl}} = 800000\text{ g bulk chemical}$	(A.10)
Volume of bulk chemical required per recovery clean	$= 800000\text{ g} \cdot \frac{1\text{ mL}}{1.175\text{ g}} \cdot \frac{1\text{ L}}{1000\text{ mL}} = 680\text{ L bulk chemical}$	(A.11)

5.0 REFERENCE

Water Environment Federation; American Society of Civil Engineers; Environmental and Water Resources Institute (2009) *Design of Municipal Wastewater Treatment Plants,* 5th ed.; WEF Manual of Practice No. 8; ASCE Manual of Practice and Report on Engineering No. 76; McGraw-Hill: New York.

Appendix B

Glossary

activated sludge—The biologically active solids in the activated sludge process.

activated sludge process—A biological treatment process where flocculated microorganisms are suspended in wastewater to facilitate the removal of organic matter and nutrients.

aerobic—Condition where oxygen is the primary microbial electron acceptor.

aerobic solids retention time—The solids retention time within the aerobic portion of an activated sludge tank.

air scour—Vigorous aeration along the membrane used to reduce the physical deposition of membrane foulants in the membrane cake layer.

air sparging—See *air scour*.

alpha factor—The ratio of the mass transfer coefficient in wastewater to the mass transfer coefficient in clean water.

anaerobic—Condition characterized by the absence of oxygen, nitrate, or nitrite as available electron acceptors.

anoxic—Condition where nitrate or nitrite is the primary microbial electron acceptor.

backpulse—See *backwash*.

backflush—See *backwash*.

backwash—Pumping water or chemical solution through the membrane in the reverse direction of permeate flow.

biofilm—An accumulation of microbial growth on the surface of an object.

cake fouling—Membrane fouling caused by the accumulation of colloidal and suspended material above the membrane surface.

capillary suction time (CST)—A type of static filtration test that measures the filtration rate and is widely used to measure the dewaterability of activated sludge.

chemical cleaning—The removal of membrane foulants from the membrane surface and within the membrane pores through chemical methods.

chemically enhanced backwash—See *maintenance cleaning*.

chemically reversible fouling—Fouling that can be removed using some form of chemical cleaning.

clean-in-place (CIP)—See *recovery cleaning*.

concentrate—See *retentate*.

conversion—See *recovery*.

critical flux—An operating flux above which membrane fouling occurs at an unacceptably rapid rate.

extracellular polymeric substance (EPS)—A complex solid matrix of proteins, polysaccharides, lipids, and nucleic acids surrounding the microorganisms in biofilms and activated sludge flocs.

filtrate—See *permeate*.

flux (J)—Flow per unit membrane area.

foulant—A soluble, colloidal, or particulate substance that causes membrane fouling.

gross membrane production—The total volume of water filtered through the membrane system in a given day.

hand cleaning—See *mechanical cleaning*.

heterotrophic bacteria—A type of bacteria that derives its cellular carbon from organic carbon.

hydraulic retention time (HRT)—The length of time that a given hydraulic loading of wastewater or solids will be retained in a pipe, reactor, unit process, or facility.

instantaneous flux—The amount of water filtered through a collection of membranes at any given moment divided by the membrane area in service.

irrecoverable fouling—See *irreversible fouling*.

irreversible fouling—Fouling that is permanent and cannot be removed physically, hydraulically, or chemically.

large membrane subunit—An assembly of small membrane subunits packaged together in a support structure and connected to a common permeate manifold.

localized dewatering—The excessive accumulation of solids within the membrane module.

manual cleaning—See *mechanical cleaning*.

maintenance cleaning—Short-duration chemical cleaning.

mean cell residence time—See *solids retention time*.

mechanical cleaning—The removal of membrane foulants, rages, and/or debris by physically cleaning the membranes by hand or by spraying down with water.

membrane—A selective barrier that permits the passage of some components, but not others.

membrane array—See *membrane train*.

membrane bank—See *large membrane subunit*.

membrane cassette—See *large membrane subunit*.

membrane cleaning—The mitigation of membrane fouling through physical or chemical methods.

membrane element—See *small membrane subunit*.

membrane fouling—The gradual accumulation of contaminants on the membrane surface or within the porous membrane structure that inhibits the passage of water, thus decreasing membrane productivity.

membrane module—See *small membrane subunit*.

membrane panel—See *small membrane subunit*.

membrane plate—See *small membrane subunit*.

membrane rack—See *large membrane subunit*.

membrane system—The sum of all the membrane trains plus the ancillary equipment required for membrane cleaning and operation.

membrane train—An assembly of large membrane subunits sharing common permeate piping, controls, and pump.

membrane unit—See *large membrane subunit.*

microfiltration—A pressure-driven membrane filtration process that typically uses membranes with a pore-size range of approximately 0.1 to 0.4 micrometers (μm).

mixed liquor—The mixture of activated sludge and wastewater being treated.

mixed liquor recycle (MLR)—A process stream in the activated sludge process where mixed liquor is pumped from the end of the bioreactor to upstream of the bioreactor.

mixed liquor suspended solids (MLSS)—The total suspended solids concentration of the mixed liquor.

mixed liquor volatile suspended solids (MLVSS)—The volatile fraction of the mixed liquor suspended solids.

monod equation—An equation typically used in wastewater treatment models to describe the kinetics of biological growth. Identical in form to the Michaelis–Menten equation often referred to in industrial applications.

net membrane production—The total volume of water filtered through the membrane system in a given day minus any losses because of physical or chemical cleaning requirements.

nitrifying bacteria—A type of bacteria that can oxidize ammonia and nitrite.

oxygen consumption rate—See *oxygen uptake rate.*

oxygen uptake rate (OUR)—The mass of oxygen required for biological oxidation of wastewater per unit reactor volume per time.

permeability—The ratio of flux to transmembrane pressure.

permeate—Treated water that has passed through the membrane.

permeate velocity—See *flux.*

physical cleaning—The removal of membrane foulants from the membrane surface and within the membrane pores through physical methods such as air scouring or backwashing.

physically reversible fouling—Fouling that is temporary and can be removed using some form of physical cleaning.

poly-phosphate accumulating organism (PAO)—A type of bacteria with a metabolism that facilitates biological removal of phosphorus from wastewater.

pore size—The size of the openings in a porous membrane expressed either as nominal (average) or absolute (maximum), typically in terms of microns.

pore blocking—Membrane fouling caused by the accumulation of soluble and/or colloidal material in the membrane pores.

product water—See *permeate.*

recovery cleaning—Long-duration chemical cleaning.

relaxation—The cessation of membrane production while the air scour remains active.

resistance (R)—The measurement of the degree to which the flow of water is impeded by a membrane material or fouling. The inverse of permeability.

respiration rate—See *specific oxygen uptake rate.*

retentate—Water and other materials retained by the membrane.

retention time—See *hydraulic retention time.*

return activated sludge (RAS)—Activated sludge that is returned to the beginning of the activated sludge process to mix with raw or primary settled wastewater.

return sludge—See *return activated sludge.*

sludge age—See *solids retention time.*

sludging—See *localized dewatering.*

small membrane subunit—The smallest assembly of filtration equipment that is designed to be removed or replaced as an integral piece.

solids retention time (SRT)—The average amount of time a microorganism is retained in the activated sludge process, calculated as the mass of solids in the activated sludge process divided by the mass of solids wasted per day.

soluble microbial product (SMP)—Soluble or colloidal compounds of microbial origin that are produced during the biological treatment process.

sorption—Membrane fouling caused by physical or chemical binding of soluble membrane foulants to the membrane surface.

specific flux—See *permeability*.

specific oxygen uptake rate (SOUR)—The mass of oxygen required for biological oxidation of wastewater per mass of volatile suspended solids per time.

sustainable flux—The operating flux at which the membrane permeability decreases at an acceptable rate given the physical and/or chemical cleaning methods applied.

transmembrane pressure (TMP)—The pressure difference across the membrane, or the driving force required to achieve a given flux.

ultrafiltration—A pressure or vacuum-driven membrane filtration process that typically uses membranes with a pore-size range of approximately 0.01 to 0.05 micrometers (μm).

waste activated sludge (WAS)—Excess activated sludge that is removed from the activated sludge process.

Index